To Peter
With compliments

Kostas - George
28. 9. '89.

METHODOLOGICAL ASPECTS OF THE DEVELOPMENT OF LOW TEMPERATURE PHYSICS 1881–1956

SCIENCE AND PHILOSOPHY

This series has been established as a forum for contemporary analysis of philosophical problems which arise in connection with the construction of theories in the physical and the biological sciences. Contributions will not place particular emphasis on any one school of philosophical thought. However, they will reflect the belief that the philosophy of science must be firmly rooted in an examination of actual scientific practice. Thus, the volumes in this series will include or depend significantly upon an analysis of the history of science, recent or past. The Editors welcome contributions from scientists as well as from philosophers and historians of science.

KOSTAS GAVROGLU and YORGOS GOUDAROULIS

METHODOLOGICAL ASPECTS OF THE DEVELOPMENT OF LOW TEMPERATURE PHYSICS 1881–1956: CONCEPTS OUT OF CONTEXT(S)

KLUWER ACADEMIC PUBLISHERS
DORDRECHT / BOSTON / LONDON

Library of Congress Cataloging-in-Publication Data

Gavroglu, Kōstas.
Methodological aspects of the development of low temperature physics 1881-1956 / Kostas Gavroglu, Yorgos Goudaroulis.
p. cm. -- (Science and Philosophy)
Bibliography: p.
ISBN 9024736994
1. Low temperature--History. 2. Superconductivity--History. 3. Superfluidity--History. I. Goudaroulis, Yorgos. II. Title. III. Series.
QC278.G38
543'.56'09--dc19 88-3798

Published by Kluwer Academic Publishers,
P.O. Box 17, 3300 AA Dordrecht, The Netherlands.

Kluwer Academic Publishers incorporates
the publishing programmes of
D. Reidel, Martinus Nijhoff, Dr W. Junk and MTP Press.

Sold and distributed in the U.S.A. and Canada
by Kluwer Academic Publishers,
101 Philip Drive, Norwell, MA 02061, U.S.A.

In all other countries, sold and distributed
by Kluwer Academic Publishers Group,
P.O. Box 322, 3300 AH Dordrecht, The Netherlands.

Printed in The Netherlands

ἐάν μή ἔλπηται,
ἀνέλπιστον οὐχ ἐξευρήσει,
ἀνεξερεύνητον ἐόν καί ἄπορον.

Ἡράκλειτος

He who does not expect the
unexpected will not detect it:
for him it will remain
undetectable, and unapproachable

Heraclitus

Table of Contents

Preface

This book is primarily about the methodological questions involved in attempts to understand two of the most peculiar phenomena in physics, both occurring at the lowest of temperatures. Superconductivity (the disappearance of electrical resistance) and superfluidity (the total absence of viscosity in liquid helium) are not merely peculiar in their own right. Being the only *macroscopic* quantum phenomena they also manifest a sudden and dramatic change even in those properties which have been amply used within the classical framework and which were thought to be fully understood after the advent of quantum theory.

A few years ago we set ourselves the task of carrying out a methodological study of the "most peculiar" phenomena in physics and trying to understand the process by which an observed (rather than predicted) new phenomenon gets "translated" into a physical problem. We thought the best way of deciding which phenomena to choose was to rely on our intuitive notion about the "degrees of peculiarity" developed, no doubt, during the past ten years of active research in theoretical atomic and elementary particle physics. While the merits of the different candidates were compared, we were amazed to realize that neither the phenomena of the very small nor those of the very large could compete with the phenomena of the very cold. These were truly remarkable phenomena if for no other reason than for the difficulties encountered in merely describing them. For a mere description of superconductivity and superfluidity comes into conflict with the definitional meaning of the terms being used. How is it possible, for example, to have zero electrical resistance at any finite temperature, except possibly at absolute zero, if the very notion of electrical resistance is associated with a measure of hindrance inherent in the structure of the conducting substances themselves? How is it possible to find results differing by a factor of a million when the viscocity of liquid helium is measured by the two different yet perfectly equivalent methods — methods which themselves have been determined by the very definition of viscocity itself? Notwithstanding the recent startling developments in "high temperature" superconductivity, a satisfactory explanation of superconductivity and superfluidity was provided by the mid- and late-fifties, coinciding, in a way, with the period which was to become quite decisive for the developments in philosophy of science.

It was during this period that Hanson's work inaugurated a re-examination of Reichenbach's distinction between "context of discovery" and "context of justification", and there was a revival of interest among philosophers of science for (re-)examining the "problem of discovery". Impressive studies of historical cases, especially of physics, accompanied the discussions about the more theoretical aspects of the "problem of discovery", and such investigations became one of the main concerns of philosophy of science — even more so after Thomas Kuhn published *The Structure of Scientific Revolutions.* The questions raised in the earlier works of Karl Popper, together with those in the writings of Thomas Kuhn, Imre Lakatos, Paul Feyerabend, and also of Joseph Agassi, Norwood Russel Hanson, Gerald Holton, Dudley Shapere, Stephen Toulmin and others concerning the implications of history of science for philosophy of science, created an especially favourable climate inspiring scientists to become active in this area of scholarship. Our overall approach in attempting to examine analytically both superconductivity and superfluidity is one which emphasizes the methodology — rather than the logic — of discovery. It is a view which has been convincingly expounded by Gary Gutting and Thomas Nickles.

This book is also an attempt to discuss a series of questions that are not being given the attention they deserve within the overall problematique of the studies of discovery. The central issue of our arguments is the methodological questions involved in the "translation" of an observed new and unexpected phenomenon into a physical problem and the ensuing process of solution for such a class of problems. Not every well-formulated problem is a problem which can be eventually resolved just because it is well posed. The discussion, in fact, of whether a problem is well posed or not, presupposes that among the multitude of (available) problems related to a specific unexplained phenomenon, a choice has already been made of *that* problem whose solution, the scientific community believes, will eventually lead to the explanation of the phenomenon. It should, thus, be emphasized that not all well-formulated problems referring to a particular phenomenon are of the same methodological status, since a phenomenon, more often than not, is "made up" of many properties each one of which, depending on which problem is chosen, has distinctly different methodological roles in the strategy to be followed for eventually explaining the phenomenon.

It is *not* our aim to use the two case studies in order to provide evidence and arguments in favour of a particular theory of scientific change. Such a task would have expressed an intolerable naiveté for, if it is anything that the examination of these cases displays, it is the futility of such a pursuit. It is not, in fact, too exaggerating to claim that given enough patience and a flair for details, it is possible to use any incident in the history of science *both* to "prove" and to "disprove" a given theory of scientific change so far as its *macroscopic* viewpoint is concerned. That, however, is not the case when it comes to the testable claims of many major theories of scientific change. A comprehensive taxonomy of these claims has recently been published by Larry Laudan and his colleagues, and their work presents various theories of scientific change, in a manner amenable to testing them *vis-à-vis* the history of science. We hope that our work will contribute towards the further clarification of at least some testable claims of these theories.

A note of caution lest the title of the book give the impression that we are delivering more than we actually do. Despite our tracing the developments at the Leiden Cryogenic Laboratory before the discovery of superconductivity, and, then, the theoretical and experimental developments in superconductivity and superfluidity, we are not claiming that we are writing even part of the history of low temperature physics. We would like to hope that the clarification we attempt of a series of methodological questions (and of certain historical ones as well) could, at least partly, clear the ground for writing the history of low temperature physics. The history of low temperature physics is a particularly complex undertaking and it spans a period of over two hundred years. Whether it was the liquefaction of gases, the developments in thermometry, the measurement of physical parameters at ever lower temperatures or the discovery of new phenomena, the successes of low temperature physics closely followed the crucial developments in molecular physics first, and of quantum mechanics later. And, significantly it gave rise to the refrigeration industry with the drastic changes it brought to economic development and everyday life.

Part I of the book starts with a classification of "new phenomena" and they are divided into two large classes, each being subdivided into further categories. One class is made up of phenomena that are predicted by a theory or explicable by one, and the second comprises what we call the unexpected phenomena. It is, then, claimed that the right problem (the problem whose solution does eventually provide an explanation of the phenomenon studied) is a problem arising out of a paradoxical situation, itself created when there is at least one statement allowed to be posed by the descriptive language of a specific theory such as to reveal that the new phenomenon, as it is translated into this descriptive language, is irreconcilable with the concepts and mechanisms of such a theory. For the solution of this class of problems we propose a process we call "concepts out of context(s)". It is a process emphasizing the continuous metamorphosis of concepts due to their (de-)contextualization: When a concept is chosen for an explanation of an unexpected phenomenon there is always a change in its meaning since it is now in a context different from the one out of which it was initially derived. "Concepts out of context(s)" expresses the dynamic signified by the (at least) dual meaning of "out of" where concepts are first derived out of their (original) context, and yet they find themselves in a state away from what is familiar with respect to their (original) context, eventually acquiring a relatively autonomous status. We are, therefore, interested both in the meanings concepts have because of their place in the conceptual hierarchical structure of a theory and also in the meaning they acquire due to their role in the specific process by which they were used in the attempts to find an explanation to the new and unexpected phenomenon. Thus we are interested in the meaning concepts acquire due to their place in the contexture. It is the contextural character of their acquired meaning which gives these concepts an excess (meaning) content, something which, in this sense, does not exist for concepts whose meaning is only due to the conceptual hierarchical structure of the theoretical framework.

Part II presents in a relatively detailed manner some of the developments in low temperature physics. No study of any aspect of these developments during the 20th century can be successful without appreciating the role of the Physical Laboratory of the University of Leiden and the structure of the research conducted there till about the beginning of World War II. Surprisingly, a methodological study of the developments in superconductivity and superfluidity also relies very heavily on examining closely Leiden's "physics culture", the style of work there and, above all, the long term research strategies drawn out since the end of the nineteenth century. No person was more influential for shaping all these than Heike Kamerlingh Onnes, who became professor of the first chair of experimental physics in Holland in 1882, and who for the following forty years was involved with every major development in low temperature physics. It is a great pity that apart from some illuminating articles there is no comprehensive scientific biography of H. Kamerlingh Onnes. In Chapter 2 we concentrate on two activities of H. Kamerlingh Onnes and his early work at Leiden, and which present a tremendous interest from a methodological point of view. These are the researches on the equation of state which led to the liquefaction of helium, and the researches in magnetism which continued long after Kamerlingh Onnes' death. The liquefaction of helium by Kamerlingh Onnes was not only an achievement of improved instrumentation and new cryogenic techniques. It needed a thorough understanding of molecular physics, and, maybe, nothing was as effective for such an understanding and the long range planning of experiments in Leiden, than a theorem about the law of corresponding states proved by H. Kamerlingh Onnes in 1881—the year we choose to start our narrative.

Both the experimental and theoretical developments in superconductivity and superfluidity are analytically presented in Chapter 3 and Chapter 4. The presentation aims at providing the necessary material in order to appreciate the overall developments which eventually contributed to the successful explanation of these two phenomena 46 and 15 years respectively after they were first observed. Therefore, these two chapters are self-sufficient, and without being exhaustive historical accounts they can be read by those interested only in the historical developments. However, this material is necessary for Part III where we attempt to "read" the developments in superconductivity and superfluidity "our way", looking at the history of these two cases through our methodological proposals discussed in Part I.

Acknowledgements

The discussions we have had with many colleagues and their continuing support contributed decisively to having a work more complete in its details and less confusing in its proposals. We are sincerely grateful to them all.

Nancy Nersessian of Princeton University encouraged us from the beginning to pursue this study and her comments throughout the period we have been preparing the manuscript were extremely helpful.

Aristidis Baltas of the National Technical University of Greece, was always willing to have us think aloud, and he read and reread extensive parts of the manuscript.

Robert Cohen of Boston University, Peter Clark of St. Andrews University, Vaso Kindi of the National Technical University of Greece, Ulysses Carlos Moulines of Bielefeld University, Thomas Nickles of the University of Nevada, Nicolas Rescher of the University of Pittsburg made extensive comments on Part I.

Three protagonists in the developments of low temperature physics helped us greatly. A long interview with Professor H. B. G. Casimir and a meeting with Professor J. van der Handel clarified many aspects of low temperature research at Leiden before the Second World War. Extended conversations with Professor L. Tisza of MIT made us appreciate many subtle aspects of the theoretical developments in superfluidity and especially of the two-fluid model which he first proposed in 1938 and which retained its remarkable heuristic strength until the advent of the correct microscopic theory in the early fifties. We also thank him for making available to us letters by F. London which are not in the London archive at Duke University.

Our visit and stay at the Kamerlingh Onnes Laboratorium of the University of Leiden in order to study the archives of H. Kamerlingh Onnes, W. H. Keesom, W. J. De Haas and C. J. Gorter was made possible through grants by the Greek and Dutch Governments. This stay would not have been as enjoyable or as productive if it were not for Dr. A. J. van Duyneveldt and Professor R. de Bruyn Ouboter who did their utmost while we were there and who were always available to discuss at length any of our queries. We also want to thank Dr. Durieux of the University of Leiden, Dr. J. A. Geurst of the Delft University of Technology and Dr. S. Engelsman of the Boerhaave Museum for his help in the examination of part of Kamerlingh Onnes' archives.

We would like to express our thanks to the many people who helped us to gain access to material from the archives of Fritz London at Duke University, Heinz London at Bristol University and the taped interviews at the Niels Bohr Library of the American Physical Society in New York.

We are glad to acknowledge the warm interest and cooperation of Mrs. A. Kuipers of Kluwer academic publishers.

During the preparation of the book we stayed for extended periods at Imperial College, London; University of Tampere, Finland; Cambridge University; and the International Center for Theoretical Physics at Trieste, Italy, and we acknowledge with gratitude the willingness of our colleagues to have us in their departments.

KOSTAS GAVROGLU,
Department of Physics,
National Technical University,
157 73 Athens, Greece.

YORGOS GOUDAROULIS
Physics Division,
School of Technology,
Aristotle University of
Thessaloniki,
Thessaloniki, Greece.

PART I

The how

The paradox reveals the reality to us.
Whoever faces the paradox is also exposed to the dangers of reality.
F. Durenmatt (in his play *Die Physiker*)

We must not be shackled all the time by the desire to fit what at each moment is accepted as experimental reality.
C. N. Yang

CHAPTER 1

"Translating" unexpected phenomena into the right physical problems

1.1. Preliminaries

Low temperature physics is known primarily for its "peculiar" phenomena, for its elaborate technical details related to the experimental set ups used to study these phenomena, for the possibilities it offers for further understanding quantum mechanics, and for the opportunities it presents for technological applications. This area of research provides us with two unique phenomena: *Superconductivity*, first discovered in 1911 as the complete disappearance of electrical resistance of mercury at the critical temperature T_c, of about −269°C (a little over four degrees above absolute zero); *Superfluidity*, first realized in 1938 after the observation that liquid helium at a temperature, T_λ, of about −271°C (a little over two degrees above absolute zero) has extremely low viscosity and can pass through the narrowest capillaries. As it happens, hardly any attention has been paid to the extremely intriguing methodological issues suggested by the various attempts to find an explanation for these unique *macroscopic* quantum phenomena.

Nearly all aspects of these amazing phenomena were totally unanticipated when they were first observed experimentally. Before proceeding, in Part II, with a detailed presentation of both the experimental and theoretical developments in superconductivity and superfluidity, we shall first discuss some of the methodological questions involved in the "translation" of an observed new phenomenon into a physical problem and its eventual solution. Although most of our comments have been derived from our study of the development of low temperature physics, our overall proposal does, we believe, have a more general validity. We shall attempt to argue our case by:

- Discussing the various difficulties in observing new phenomena and differentiating among the many kinds of new phenomena;
- Establishing criteria for recognizing the new phenomena which are unexpected;
- Defining what we mean by the emergence of a paradoxical situation when the

observed unexpected phenomenon is expressed by the descriptive language of the dominant explanatory schema;

- Explaining how a paradoxical situation leads to the formulation of a physical problem, realizing, at the same time, that the difficulty resides in finding the *right* problem, and not just in formulating correctly *any* problem;
- Articulating a process which we refer to as constructing "concepts out of context(s)" and which appears to be followed in the resolution of this class of problems.

The following points "codify" some of the relatively unique characteristics of superconductivity and superfluidity and signify the directions of our methodological study:

i. The infinite electrical conductivity of mercury wire and the discontinuity in the change of the various thermodynamic parameters of liquid helium were the two phenomena which heralded the beginning of research programs in superconductivity and superfluidity, respectively.

Superconductivity and superfluidity exhibit a characteristic difference the overall consequences of which have resulted in the multitude of methodological trends present in low temperature physics. When superconductivity was first discovered, the emphasis was to find an explanatory schema in order to account for the observed infinite conductivity. On the other hand, the phenomenon of superfluidity was not associated with a single property and, in contrast to superconductivity, it presented a difficulty as to which of the many "peculiarities" of supercooled helium constituted its most dominant characteristic so that the understanding of this property would, in turn, lead to an understanding of the rest: the extremely low viscosity, the discontinuous change of the specific heat at the λ-point, the extremely high thermal conductivity, the "fountain" effect, or the fact that helium, under its own pressure, remains liquid down to 0°K. In the case of superconductivity the "final" formulation of what we shall call the "right" problem was achieved 22 years after the initial observation of the phenomenon. During all this period, the people working to understand infinite electrical conductivity were, in effect, working on the "wrong" problem. By contrast, part of the difficulty with superfluidity was the inability of formulating a single problem, since understanding the phenomenon involved the formulation of many "right problems".

ii. The methodological implications of the relationship between theory and experiment become absolutely crucial in comprehending the overall novelties of this area of physics. There are the "landmark" cases where experimental results "legitimize" the use of theoretical techniques which were otherwise forbidden. There are also heuristic arguments supporting specific theoretical approaches and lead to a revaluation of already existing experimental data and the planning of new experiments. Furthermore, during Kamerlingh Onnes' directorship of the Physical Laboratory of the University of Leiden, a specific attitude dominated the (intentional) relationship of theory to experiment and to experimentation generally. Eventually, and as a result of this specific attitude about the relationship between theory and experiment, a characteristic "physics culture" (which we shall call "sophisticated phenomenology") emerged at Leiden.

iii. Superconductivity and superfluidity are macroscopic quantum phenomena. They are not simply the macroscopic manifestations of quantum mechanics which, for example, are witnessed in other phenomena (radioactivity, spectra, etc.). Here we are referring to phenomena where one will have treat quantum mechanically the macroscopic physical system itself and not just its microscopic mechanisms. Hence, in attempting to understand these phenomena, one is confronted primarily with a *conceptual challenge*, since the appropriate concepts and their related framework became a test of the extent of the validity of quantum mechanics.

iv. There are two distinct methodological approaches to formulating an explanation which will encompass both superconductivity and superfluidity. These strategies are used interchangeably in the various research programs, depending on the experimental results, the direction of theoretical developments, and the "problematique" within the scientific community. One strategy is to use a generalized concept of "flow" (both electrical and fluid) and to consider the phenomena as states of aggregates. The other strategy is to use the technique of particle interactions and continue the program by introducing various concepts which express the specificity of these interactions. The dilemma posed by the problem of whether to use techniques involving collective behaviour, and the physical parameters characterising collectiveness, or the particle-interaction techniques is not confined to low temperature physics. The choice between the two approaches is not primarily determined by the intricacies of the mathematical techniques applied or devised for such cases of collective behaviour. The criteria used for such a choice are also determined by metatheoretical and metaphysical considerations and express specific methodological preferences concerning the overall development of particular research programmes.

The methodological considerations involved in formulating (and reformulating) observed new phenomena into problems amenable to solution, which, in turn, are accepted by the scientific community as themselves problems to be systematically and critically examined, have not been properly investigated. Almost all the main approaches to philosophy of science regard a problem as given or at least as not being their central concern. This is so, independent of the difference among the various approaches concerning the methodological status of the problem solving process. Is the "translation" of a new phenomenon into a physical problem really as straightforward a process as seems to be suggested by the reconstruction of various historical cases? Why is it that these reconstructions regard this "transition" from the new phenomenon into a physical problem as one of the least problematic aspects of their overall methodological approach? The questions we are raising are only partially related to the difficulties associated with a badly formulated or ill-posed problem. What we shall discuss instead is the process of translating a new phenomenon into the right problem and in such a language as to suggest the possible strategies that have to be followed for its resolution. A new phenomenon needs to be stated as a definite physical problem, and not merely as a difficulty or an anomaly, *always bearing in mind that the study of the attempts to solve a problem cannot be considered as being independent from the study of the attempts to formulate it.* We hope to show that being able to articulate the constraints inherent in the formulation of the right problem gives us an insight into the heuristic of the subsequent attempts to solve the problem. Such a "translation" must further ensure the consensus of the scientific community that the specific problem is what should, in fact, acquire a primary role in its research activity. This consensus is necessary since it is not always the case that every well formulated problem is a problem that attracts the attention of the entire community.

Let us make three introductory comments:

i. Nearly every discussion concerning the "methodology of discovery" emphasizes the problems related to more or less major theoretical advances. Nevertheless, the details of various "everyday" and "non-glamorous" experimental investigations, and the physical problems to which they lead, also deserve some attention, since their implication for theory construction cannot really be negleted.

    Part of our discussion has to do with the Peirce — Hanson[1] concept of a "surprising phenomenon" and with Kuhn's[2] thesis that anomalies are "violations of expectations".

    Hanson, interestingly, confines his "surprising phenomena" to those which give rise to an "astonishment (which) may consist in the fact that p is at variance with accepted theories. What is important here is *that* the phenomena are encountered as anomalous, not *why* they are so regarded".[3] Hanson assumes that his surprising phenomena are readily turned into well-posed problems, so that these phenomena will cease to be surprising when one adopts a suitable theoretical hypothesis H. But this assumption can be questioned on two grounds. First, on historical grounds, there are many cases where the "translation" of the surprising phenomenon into a well-posed problem is very far from being a straightforward process. Second, on methodological grounds, one is also obliged to understand *why* the astonishing phenomena are so regarded; for example, what are the criteria for assessing "unexpectedness"? can one talk about "degrees of unexpectedness"? and so on. If one pays attention only to the fact that a phenomenon is surprising (and not to why it is surprising) then, sooner of later, there remain only two ways to find out whether one is confronted with surprising phenomena or not. Candidates for this class of phenomena will be among those which either do not agree with the predictions of a theory or are observed for the first time and cannot readily be explained by the existing theory. These are indeed surprising phenomena, but neglecting *why* a phenomenon is surprising overlooks another very important procedure for judging whether we are confronted with such a phenomenon: There are instances where an *agreement* with the theoretical predictions, followed by a reconsideration of the theoretical inputs and the interpretative aspects of a theory, *itself* becomes a manifestation of a new phenomenon. Thus, the observed agreement can lead to the overthrow of the schema which, at first sight, was thought to be corroborated by this phenomenon.

There are, no doubt, many cases where the expression of a surprising phenomenon in the descriptive language of the dominant theory turns it into a well-formulated problem. There are, however, many *other* cases where such a formulation is a much more complex procedure, and it is these cases which will interest us here. By examining the transition from surprising phenomena to well-posed physical problems and their resolution, we shall try to show that the answer to *why* a phenomenon is regarded as astonishing becomes so significant in our approach.

ii. In the following discussion, we shall primarily inquire into a set of methodological issues which precede the question as to whether we have a well-posed problem. These are the methodological problems involved in the "translation" of an observed (new) phenomenon into a physical problem. Thus, our work also involves those cases where the failure of the scientific community to resolve a problem is *not* necessarily due to its bad formulation. A problem may not be solved because what has been "translated" into a physical problem may, in fact, be a peculiar phenomenon, which does not satisfy a set of criteria so as to be "appropriate" for such a translation. *Hence, a problem could be well formulated, but it may not be the right problem towards which the scientific community directs its researches. And, of course, after a theory is developed as a result of solving the right problem, the theory has to be able to resolve the initially well formulated — but wrong — problem as well.* The development of low temperature physics provides some extremely characteristic cases of this sort: cases which, even in their generality, have been neglected in the discussions concerning the "context of discovery".

iii. It is undoubtedly the case that many of the relatively recent accounts of scientific change that have emerged from a criticism of logical positivism regard the scientific community, as well as the social factors which are extraneous to the community, as factors whose influence on the development of science cannot be considered negligible perturbations to an otherwise exclusively intellectual activity. We do not intend to enter into a discussion about the relationship between philosophy of science and sociology of science but wish to emphasize that no study of the "context of discovery" within the current problematic in the philosophy of science can do justice to the various proposals, if one does not make explicit the specific social factors within the set of proposals used to understand the development of science. For it is important to realize that what are commonly referred to as social factors, when considered from the point of view of philosophy of science and the specific schemata one is working with, do not have a "model independent" character. In each case, it is necessary to articulate those aspects of the philosophical-methodological schema that make it possible to signify the *particularity* of the "social factors" *vis-à-vis* the specific schema.

For our purposes, these factors can be expressed in the following manner:

a. The difference between, on the one hand, the experimental and theoretical difficulties that result from the investigations of a particular theory, and, on the other, the way these difficulties become "problems for solution" and lead to a *consensus* within the community of physicists that *these are indeed* the problems to be systematically investigated;
b. The totality of questions addressed to the theory that aims at expanding the limits of applicability of the interpretative ability of the existing theory;
c. The fact that these questions are posed within a context influenced by prejudices in favor of the claims of the existing theory. These prejudices, by affecting the kind of questions we address "towards" the theory, restrain the "mission" of these questions, since such a mission presupposes doubting certain ontological claims of the existing theory.[4]

1.2. A taxonomy of the phenomena

Most of the discussions about taxonomic schemata are basically concerned with clarifying the inherent ambiguities as to how the taxonomic criteria are defined. The schema we propose is by no means free of these ambiguities. Nevertheless, it becomes possible to have a useful categorization of new phenomena, if we choose to divide them into various types depending on their relation to the dominant theory.

i. Phenomena predicted or explained by a theory

This class includes five types as described below:

Type a phenomena that are predicted by a specific theory or model and, when observed, fit the theoretical prediction exactly. No modifications to the theory are needed to provide an explanation of any aspect of them. This class of phenomena is extremely important for providing corroborative evidence for a newly-proposed theory or model and usually comprises those new phenomena predicted right after the formulation of the theory. Such, for example, is the case with the discovery of the Ω^- particle, when the proposed group-theoretical approach for the classification of elementary particles in the early sixties predicted a hitherto unobserved particle with the quantum numbers that Ω^- later turned out to have.

Type b phenomena that are the result of a predicted behavior whose observation corroborates a theory or a model; without, however, the theory or the model being able to provide a quantitative account of the observed specific behavior as well. This account may be achieved either after a relatively straightforward development of the initial theory or it may be the result of an involved and major breakthrough.

Let us consider two examples:

When Lee and Yang in 1956 proposed that parity may not be conserved in the weak interactions to explain the so called "$\tau - \theta$ puzzle", they proposed a type of experiment in which an asymmetrical distribution in space of outgoing electrons could be examined. The experiments did, in fact, record such an asymmetry. Their proposal, however, could not account for the magnitude of the asymmetry. The observed "maximal asymmetry" was explained a couple of years later in a relatively straightforward modification of the original Fermi interaction for the weak decays.[5]

A similar situation, which involved far more dramatic developments, was the predictions of Einstein's paper in 1911 in which he investigated the implications of the principle of equivalence — "the happiest thought of my life". In that paper, he predicted two totally new phenomena: the bending of light when it passes near a massive body and the "acceleration" of falling photons. No experiments were performed at the time but, if such experiments had been performed, these predicted phenomena would have been observed, though it would not have been possible for the magnitude of some of the observed quantities to be accounted for by available theory. It was only with the full development of the general theory of relativity in 1916 that the correct quantitative description of these phenomena was obtained.

This specific example allows us to make a crucial point concerning new phenomena. The 1911 paper does, in fact, predict the new phenomenon of the bending of light by the gravitational field, resulting directly from the principle of equivalence which concerns the local equivalence of the inertial and gravitational forces. The new calculation in the 1916 paper, however, in a way predicts the *same* effect as a *new* phenomenon: The amount of the bending was now calculated to be double the value found by calculation using only the principle of equivalence. The bending was now due to the form of the geodesics, i.e., it was due to the effect of the gravitational field on space-time.

Hence a new phenomenon is not "new" only because a theory or a model predicts it for the first time. The same phenomenon may also be predicted by another theory (or by a developed form of the initial theory) it may be due to a totally new mechanism, and calculations may give different values for the measurable quantities. Thus, the new phenomena should also be looked for in the quantitative differentiations, since these differentiations may be due to totally new mechanisms and effects.

Type c phenomena that are not predicted by any theory. When observed, their explanation can be achieved by changing the interpretation of a specific theory. A characteristic example is the discovery of the positron. This phenomenon was explained by changing the interpretation of the solutions of the Dirac equation and adopting the point of view that the double solutions of the equation are solutions expressing a new class of elementary particles, i.e. the antiparticles.

Type d phenomena that are neither predicted by a theory as such nor searched for as corroborative instances of the theory. These are phenomena that seem to be "hidden" in the theory itself (or its logical consequences) and surface only during its normal development. One may claim that what we have here is a theory predicting new phenomena and indeed such may be the case, apart from the fact that the role of these phenomena is not so dominant in the corroborative procedures for the theory. Such is, for example, the case with the Coriolis force, when G. Coriolis in the 19th century analysed the phenomenon — without being under the "pressure" of any experimental anomalies — and thus revealed a new aspect of Newtonian mechanics.

Type e phenomena that are not predicted by theory. They are first experimentally observed, but their explanation cannot be readily found within the existing theory. The puzzling situation they create is finally resolved by a further refinement of the dominant theory. If elaboration and refinement of the theory results in actually explaining the phenomenon, then this counts as considerable confirmation of that theory over its rivals. This type also includes those phenomena which, because of the details of the proposed explanation, are used as probes for the study of various properties. In this respect, they expand the domain of the possible phenomena that we can now investigate. Examples of such new phenomena are the Auger effect, the Mössbauer effect, the Čerenkov effect, the de Haas — van Alphen effect.

ii. The "unexpected phenomena"

This class includes three types as described below:

Type a phenomena that are observed during the normal development of a theory and are considered as unexpected because they, initially at least, do not seem to fit the "overall philosophy" of the specific theory. What is interesting is that it is almost always possible to "read" the theory in such a manner as to account for the qualitative features which make the observed phenomenon an unexpected phenomenon in the first place. Even if the qualitative features of the observed phenomenon can be accounted for, its characterization as an unexpected phenomenon remains valid.

When a satisfactory explanatory theory is eventually proposed, it turns out that this particular phenomenon was one of the "falsifying" instances of the specific version of the theory so far countenanced.[6] This is so despite the fact that the phenomenon may coexist with the theory for a long time and that it may not be considered by the scientific community as a disturbing factor so far as the theory's validity is concerned. Such was, for example, the case with the excess in the precession of the perihelion of Mercury. Newtonian mechanics was able to provide a qualitative account for the phenomenon (by constructing elaborate models with matter responsible for the observed excess). There were even people who believed the excess to be an indication that gravitational forces may not vary exactly as the inverse square of the distance. The scientific community, however, *as a whole*, was not really bothered by the new phenomenon which was *not* considered to question the validity of Newtonian mechanics. This, as it turned out, was a wrong belief since the excess in the precession of the perihelion of Mercury was, in fact, the only phenomenon that Einstein's General Theory of Relativity, as formulated in his 1916 paper, predicts both qualitatively together with the exact magnitude of the excess. The other two (that of bending of light and the acceleration of the falling photons) are predicted as new phenomena in a straightforward manner by just the equivalence principle. The diamagnetic character of superconductors and the fact that liquid helium remains liquid under its own pressure down to absolute zero are also examples of this type of unexpected phenomena.

Type b phenomena that are recorded as unexpected because they do — or seem to — violate model- and theory-independent "principles" which are most vital both in theory construction and for the underlying ontology of these classes of theories. Such are the conservation laws which are the physical manifestations of well-defined symmetry properties of each theory, mathematically expressed via a set of transformations which in turn "mathematize" basic physical demands. The mathematical transformation, for example, which demands that a theory have an invariant form at every point of space and time between inertial frames, leads to the conservation of momentum and energy. Other invariance demands are not so "natural" and the conservation laws they lead to are not so commonly employed. Interestingly, the experimental testing of many such "laws" has been performed relatively recently, since not many members of the scientific community thought the testing of some of these laws to be a worthwhile undertaking. For such tests, it was assumed, were incapable of giving anything different from what was naturally (!) expected. It is prejudices in favour of a particular structure of nature, usually determined by conservation laws not rigorously tested, that play a significant role in "determining" our expectations.

It may, thus, be the case that an observed phenomenon may seem to violate a specific conservation law, as was the case with the electron spectrum of the beta decay, prior to the proposal for the existence of the neutrino, where there seemed to be a breakdown of energy conservation. Alternatively, a phenomenon may defy explanation if we insist on using a theory or a model where we have demanded that a particular invariance be incorporated into that theory or model. The "$\tau - \theta$ puzzle" is a good case in point: no successful explanatory schema could be devised since there was an insistence on assuming conservation of parity. A phenomenon may be unexpected because it is not possible to regard it as a manifestation of an assumed conservation law which resulted from a specific mathematical constraint imposed from the beginning on the construction of adequate theories.

Not all demands for invariance, and the ensuing "conservation" laws, are of the same ontological and methodological status. Such demands play a primarily heuristic role. We know from the outset that they are not exact and that they would be violated in a certain class of cases, though we may not know the systematic manner of their violation. Phenomena which seem to violate these latter "conservation" laws are not to be considered as unexpected since we expect a violation, but we may not anticipate the "manner" and exact magnitude of such a violation. These are cases where we have deviations from the predicted values as, for example, in the planetary motions. We expect them to happen due to pertubations and most of the time we can account for them exactly. There may also be those cases where we have used, mainly for mathematical convenience, a procedure (such as the group theoretical methods in elementary particles) that we know is idealized but usually do not know the exact form of what causes the expected deviation.

Type c phenomena are unexpected phenomena that are observed when scientists, in their attempt to construct an explanatory schema for a physical effect, are searching for any regularities, any "blueprint", which may be manifested by the extensive gathering of data that may facilitate the construction of a theory or model. These data are being gathered while varying a particular (set of) variable (s) as initial input. In these situations, there usually exists a model, a more involved schema, or the rudiments of a theory, all of which are often not capable of explaining the effects caused by these variations. Scientists search through the data they are systematically collecting for clues as to how to proceed with theory construction. The appearance of unexpected phenomena is almost always an indication that the construction of a theory for all possible values of a variable is impossible by any "extension" of the existing schema. A radical break is heralded by these phenomena, and the need to introduce new concepts is imminent.

The formulation of a satisfactory theory for electrical conduction was high on the agenda of physicists at the end of the 19th century, especially after the discovery of the electron and the development of thermodynamics. Such a theory, of course, was required to predict the values of the electrical resistance for as large a range of temperatures as possible. This was especially so for low temperatures where the effects of perturbations due to thermal disturbances were thought to be relatively negligible so that, as the temperature fell, the conduction mechanisms could be "observed" directly without interference from the thermal disturbances. There were no specific predictions for the behaviour of the resistance of conductors at low temperatures, but there were two conflicting proposals: (1) thermodynamic arguments would lead to the conclusion that since all thermal motion would cease at absolute zero the resistance to the conduction of electrons would progressively tend to zero as the temperature declined; (2) according to a proposal by Lord Kelvin, however, the resistance of a conductor would reach a minimum, would then begin to increase and would be "infinite" at absolute zero, because at that temperature the conduction electrons would "freeze" on to the atoms. Such was the situation when the resistance of very pure mercury was measured at liquid helium temperatures in 1911. The result was a truly "unexpected" phenomenon: the resistance was nearly zero in a way which could not be accommodated within the framework of either one of the proposals. The resistance dropped to very nearly zero at a temperature somewhat higher than absolute zero, and it did so abruptly as this critical value was approached.

The density, the thermal conductivity, the specific heat, and the viscosity of liquid helium all showed very unusual features, whose "articulation" within the overall theoretical framework of the molecular hypothesis proved highly problematic.

In discussing the classification of new phenomena we cannot ignore a kind of phenomena which, without strictly speaking, falling in the above categories, do indeed have a place in such a discussion.

A very unusual yet methodologically important class of phenomena are those which are "proposed" in the form of *gedankenexperimenten* and which are used to define new concepts which, in turn, play a crucial role in the development of a theoretical framework within which such a phenomenon was initially "conceived". The thought experiment with the elevators and the ray of light which led Einstein to formulate his principle of the (local) equivalence between gravitational and inertial forces did not only serve as one of the foundational principles of the General Theory of Relativity but also predicted the bending of light by the gravitational field. Here we are confronted with a new phenomenon that resulted from a specific setup where no theoretical principle forbade us from constructing such a setup. Deriving new concepts, then, becomes a relatively straightforward step, following the description of such a new phenomenon.

Carnot's analysis of the heat engine in 1824 is another example of this type of phenomenon. By examining the situation with falling water and a watewheel, Sadi Carnot concluded that the production of the motive power, whenever there is a temperature difference, must depend on both the caloric employed and the size of the temperature interval through which it falls. In his concept of reversibility, Carnot also implicitly assumed the converse of this premise: that the expenditure of motive power will return caloric from the cold body to the warm body. Carnot's analysis which was not appreciated at the time, was eventually used by Lord Kelvin in 1848 (who discovered it through the writings of Clapeyron) to define an absolute temperature scale independent of the thermometric substance. More importantly, however, Clausius in 1854 identified a physical property he called "entropy" from his study of Carnot's cycle — a development on which Boltzman based his thermodynamics. New phenomena, such as Einstein's elevator and Carnot's heat engine, are really used as probing devices to further explore the possibilities offered by a specific theoretical framework for defining new concepts.

1.3. Paradoxical situations and the "right" problems

In considering unexpected phenomena, one will necessarily have to discuss what criteria, if any, exist for assessing the "degrees of unexpectedness" of the various new phenomena. Such a discussion should not aim at the futile task of quantifying "unexpectedness" but rather at attempting to analyse the features of the context within which observed phenomena are characterised as unexpected.

We are led to expect a specific behaviour not because of what a theory could (not) predict, but because of what the implications of the extrapolated domain of the theory will have us believe. The expected behaviour (again, independently of whether it is being specifically looked for or not) is an expression of a complex of prejudices that are instrumental in both allowing for and constructing this extrapolated domain. Our expectations are aroused not only because of well-defined theoretical criteria, but also because of a series of metacriteria which have been progressively formed due to the successes of specific explanatory schemata and to repeated empirical observations.[7]

One could claim *a posteriori* that the features built into any theory, subjected as it is to the collective consciousness of the scientific community, lead to the formulation and the silent acceptance of a network of prejudices. Such acceptance is usually expressed by the unwillingness of the community to question "results" that they consider obvious or self-evident. These "results" are usually taken as the points of departure for building those schemata which fail to provide any explanations for the problematic phenomena. It would be absurd to claim that a systematic investigation of prejudices related to specific case studies can lead to an overall prescription of how to avoid them in the future. There is, however, merit in studying them both in order to render them explicit during the development of various theories and in order to understand the reasons that created them at every specific instance in this development.

To undertake such a study presupposes, and demands at the same time, that certain methodological questions be tackled. Against what criteria are the prejudices to be judged? Does this imply the acceptance of a metatheory or of a set of criteria whose validity would eventually be judged on either ideological or metaphysical grounds? A series of answers may be formulated, though all suffer from an inherent weakness in determining the "measure of criticism".[8] An answer along slightly different lines is given by the enumeration of these prejudices as judged against the later development of the theory which was supposed to have been hindered by these prejudices. Nevertheless, any successful approach to answering these questions necessarily presupposes a systematic "listing" of those historical instances where kinds of prejudices have had an especially pronounced effect. Superconductivity and superfluidity are particularly instructive cases for the study of such prejudices.

Whenever we are confronted with a new phenomenon, the initially observed surprising property is followed by (many) others all of which eventually "comprise" the phenomenon. Without taking what follows too literally, the degree of unexpectedness of a new phenomenon may be said to be "proportional" to the number of unusual properties of the phenomenon itself. Even though we shall consider the collective expression of all these properties as "composing" the new phenomenon, it is important to differentiate between the various features of these properties, since — in retrospect — not all the properties necessarily have the same methodological significance for establishing what we call "the paradoxical situation" and lead to the formulation of the relevant physical problems.

(i) *A referential* (r) property is that property a reference to which facilitates "naming" and "identifying" the phenomenon. It is a property whose connotations — mainly for historical reasons — together with the dramatic break with expected behaviour, is a more convenient expression of the phenomenon as a whole. Such is the case, for example, with the almost complete disappearance of electrical resistance in superconductivity. In superfluidity it is the abnormally low viscosity of liquid helium which allows it to pass through the narrowest capillaries.

(ii) *A significative* (s) property is that property which is more intriguing on technical and theoretical grounds and which usually defies explanation even within the rather ill-defined confines of an initially explanatory schema that attempts to find a rough qualitative account of (usually) the referential property. The destruction of superconductivity by high magnetic fields is an example of this property. In superfluidity, all of the changes in the thermodynamic parameters at λ-point without a change of phase of liquid helium provide other examples.

(iii) *A constitutive* (c) property is that property which leads to the formulation of the new phenomenon into a specific form of a physical problem which is eventually resolved by the scientific community. Two examples of this property are (1) the diamagnetic property of superconducting materials and (2) that helium remains liquid under its own pressure down to absolute zero.

In naming these three properties we are obviously talking about the different roles of the (new) properties independent of whether these roles can be expressed by one or more different properties. Perhaps in some phenomena we would have properties with "combined roles", e.g., a "referential-constitutive" (rc) property. These, then, are the *rs, rc* and *cs* properties. We may also have a single property (*rsc*) which, when further investigated and reinterpreted, may satisfy the demands of all three roles (such is the case of the equality of the inertial and gravitational mass in the way Dicke[9] analyzed the specific null experiment). What is, however, far more interesting from both a methodological and an historical point of view, is to have distinctly different properties play the different roles, as in the cases of superconductivity and superfluidity.

It is, then, this intuitive notion about the "degree of unexpectedness" which eventually induces the paradoxicality inherent in the new phenomena. What, however, is this paradoxicality? Is it merely the unexpectedness of the new phenomena expressed differently?

A "paradoxical situation" is the necessary prerequisite for formulating the new phenomenon as a well-posed physical problem that is eventually resolved by the scientific community. Becoming conscious of a new and unexpected phenomenon implies that this particular phenomenon has been at least successfully described. This *description* is achieved through the use of concepts and relations (necessarily) borrowed by the theoretical framework which, even though composed mainly by the dominant theory, is in fact something "larger" since usually the collective experience of the scientific community also influences the description of the new phenomenon.

A paradoxical situation is created when there is at least one statement posed by the descriptive language of this specific theoretical framework such as to reveal that the new phenomenon, as it is "translated" into this descriptive language, is irreconcilable with the concepts and "mechanisms" of the dominant theory. *This incompatibility is actually between two contexts. The first is the context created by the allowed extrapolations of the initial theoretical framework in the attempt to describe the observed new phenomenon in terms of the dominant theory. The second is the context formed by what is implicitly and explicitly implied by the new phenomenon when it is already expressed in terms of the descriptive language of the dominant theory.*[10]

It is this latter context which has not been adequately discussed, and its significance in the process of formulating problems has been greatly overlooked. This context is created by our theory-laden procedures of observation; once an unexpected phenomenon is described with the language of the dominant theory, it does not follow that the implications of this description are necessarily confined to the predictions, implications, or exptrapolations of the theory itself. The unexpected phenomenon acquires, through such a description, *a meaning autonomy* which, in turn, allows for those extrapolations that go beyond the theory.

When Rutherford proposed the model for the nuclear atom, after the "scattering experiment" of Geiger and Marsden, the following paradoxical situation arose. Despite the impressive agreement of the experimental measurements with the predictions of the model, the implications of the model could not be reconciled with those of the classical electrodynamic framework. At least one question revealed this paradoxical situation: given the nuclear atom, why were atoms stable? If the proposed model allows for a large "empty space" between the "central" positively-charged nucleus and the negatively-charged electrons, then the electrons will eventually spiral into the nucleus. The implications of the nuclear atom, when the model is expressed in the descriptive language of the classical framework (i.e., of classical mechanics and electromagnetism), turn out to be incompatible with the implications of the framework itself!

Another example is derived from the old quantum theory. Sommerfeld's calculations accounted for the hyperfine splittings in the spectral lines of hydrogen when the relativistic corrections due to the motion of the electrons were taken into consideration. When the hyperfine splittings of the alkaline atoms were measured, a paradoxical situation arose when an *agreement* was found between the predictions of the theory and the measured results. On second thoughts, however, this agreement turns out to be paradoxical, since in the alkaline atoms there are electrons in different orbits than those of hydrogen. Thus, the relativistic corrections on the outer electrons would be quite different since they would have different velocities; the very agreement with the experimental results cannot be always regarded as a corroborative instance in favour of the model.

The creation of a paradoxical situation is an indication that one is indeed in a position to formulate the new phenomenon into a well-posed problem for solution. The persistence of a paradoxical situation, even after the initial refinements to the theory, may also be regarded as an expression of the deadlock of any attempt to find a common ground shared by both contexts. What in effect we propose here is that whenever a problem is well formulated, it does not necessarily mean that it is also the formulation of the *right* problem. A problem need not only be well formulated, but it should also be the right problem. *It should be a problem arising out of a paradoxical situation which itself has been created during the attempts to "fit" the constitutive* (*or the rc, sc, rsc*) *properties within the existing theoretical framework.* There is no fixed temporal order to these steps, since, for example, one may become aware of the paradoxical situation after a preliminary formulation of the problem, and such awareness, in turn, may "sharpen" the formulation of the problem itself. What we, propose, then, is that the end result of a reconstruction of cases in which new theories have been motivated by the appearance of unexpected phenomena should display all of the following aspects: a constitutive (or *rc, sc, rsc*) property, a paradoxical situation, and a "good" formulation of the right problem.

Although we are interested primarily in the way new and unexpected phenomena lead to the formulation of the (right) physical problem, it is by no means the case that paradoxical situations are created *only* after establishing an incompatibility between the implications of the theoretical framework and those of the description of the newly-observed unexpected phenomenon.

There are times when models proposed right after the observation of an unexpected phenomenon do give a satisfactory account of the phenomenon, but the assumptions on which they rest are highly problematic. They may be "too" *ad hoc*, annoyingly arbitrary, or they may entail "highly unphysical" mechanisms. We are confronted, then, with another kind of paradoxical situation on which the interest and the efforts of the scientific community are concentrated. It is the incompatibility between the implications of the theoretical framework and the hypotheses involved in the proposed model. Such an incompatibility is often expressed as the ad hocness, arbitrariness or the explicitly-stated unphysicality of these hypotheses.

The emphasis now is to provide the necessary theoretical justification for the *success* of the model so that the same end result will not seem to be based on these kinds of assumptions. In other words, an attempt is made to give a phenomenological model the status of a theoretically justified explanatory mode.[11]

In such a situation our proposed methodological schema continues to be valid, since it is based on the "rise" of a paradoxical situation and its eventual disappearance through a process of "concepts out of context(s)" (with which we deal in the next section). Such a process is particularly relevant for the resolution of a paradoxical situation created by the *success* of a model which, however, is based on "problematic" hypotheses. When a paradoxical situation appears in this manner, then what the community of physicists knows is that any proposed solution will necessarily involve a reconsideration of the conceptual structure of the theoretical framework and of the meaning content and the interpretation of various concepts.

In such cases, one should keep in mind that the hypotheses are regarded as *ad hoc* and arbitrary, because of the "standards of normalcy" progressively established within a specific theoretical framework. Therefore, the originally unphysical but successful model after being theoretically justified does not appear to be "so" *ad hoc* and arbitrary (or, even, unphysical) not only because of the proposed new mechanisms, of because of new relations derived from the mathematical treatment of the implications of the first principles, but also because there would be a (radical) overthrow of the previous "standards of normalcy" mainly through the proposal of new concepts and/or the reinterpretation of existing ones. It would not be surprising to have a mathematically coherent justification for the success of a particular model and, at the same time, *demand* that a totally new set of standards and criteria of normalcy be valid, even if these look highly unphysical with respect to the existing ones. In such cases the strength of the proposed justification is so convincing that the community makes the decision to get accustomed to such new standards. Thus the paradoxical situation disappears by a reconstruction (after a process of concepts out of context(s)) of the initial situation, and the realization that there is no longer a paradoxical situation, if one *replaces* a set, at least, of the existing criteria of normalcy. Let us stress again that this involves a peculiar kind of decision on the part of the scientific community in that an "unphysical" situation becomes acceptable because it is judged that everything else involved in the proposed solution are too important to be given up just because they entail (or presuppose) an interpretation which seems unphysical.[12] It is this *willingness* of the scientific community to accept "unphysical" premises (even though other such premises create a paradoxical situation) that provides a much more complex and productive procedure for bringing about changes in our ontological beliefs, rather than merely observing the truly remarkable phenomena and waiting for their eventual explanations.

There have been some discussions about certain of the questions we are raising and the following are comments concerning those claims that are most relevant to our considerations.

In his proposed solution to the "problem" problem (i.e., the problem of developing a model of problems rich enough to account for the data arising from the history of a specific inquiry), Thomas Nickles[13] puts forth a set of "logical and conceptual requirements on problems". These criteria delineate the domain within which one should pursue the systematic investigations for this much neglected area of study. Nickles shows that both the (minimal) empiricist model and the positivist model of scientific problems cannot really meet these demands. Popper's approach, on the other hand, is not dismissed easily even though, Nickles argues, Popper cannot account for all the proposed requirements especially those that make up the evidence that the conceptual constraints which belong to the problem itself and cannot be removed — as Popper would "prefer" — to the background.

One cannot deny that, compared to the empiricist and positivist conceptions, Popper's approach is unquestionably more satisfactory. However, Popper's preoccupation with the nature of a problem is only within the context of his tetradic schema.[14] This schema does start with a problem, and Popper's aim is, as he says, to show that "the problem of understanding always turns out to be a problem about a problem"[15] while claiming at the same time that "science starts from problems".[16] Popper, in fact, goes a step further. He stipulates that "the history of science should be treated not as a history of theories, but as a history of problem situations and their modifications".[17] It also cannot be denied that Popper's discussion of "problems" (and "problem situations")[18] inaugurated a rather sophisticated approach to scientific inquiry and showed the inadequacy of the naïve positivist attitude whereby every conflict and difficulty begging for explanation was considered to be a problem. But — and this should not be overlooked — Popper's main interest is the way there is a change from one initial problem P_1, to another problem, P_2. This change is essentially brought about by that absolutely crucial step in his tetradic schema: "error elimination" aimed towards the "tentative theory", i.e. the tentative solution to the initial problem P_1.

But even in Popperian terms such a schema needs to be modified. (Gavroglu 1976; Baltas, Gavroglu 1980). The key element of the modification consists in incorporating Popper's concept of "world 3" (*W*3) in the schema and the assertion that error elimination (*EE*) is not directed towards the tentative theory (*TT*) as such, but at the *W*3 induced by this *TT*. This is so because the *TT* does provide a solution to P_1, and, hence there is no point in aiming the *EE* at the *TT*, as such, since the latter has fulfilled its role. Moreover, *EE* has a basis in world three of other theories "related with" what *TT* sets out to solve and/or world three of technology that may provide the apparatus for *EE* — all this being called the "transient world three" (*TW*3). Popper's schema can now be interpreted as follows. A tentative solution is put forward for the initial problem P_1. This *TT* generates a $W3_1$. The *EE* procedures are addressed at the implications of *TT*, that is at $W3_1$. The result of *EE* determines a $W3_2$ which is either a modified form of $W3_1$ or a synthesis of $W3_1$ and part of *TW*3. A new problem then emerges through an interaction between $W3_2$ and *TW*3.

In this approach the attempt to solve a problem is successful to the extent that such an attempt creates many other problems. Criticism, in Popper's sense, is the dominating factor both in bringing about a change from P_1 to P_2, and in creating new problems. This function becomes quite decisive and autonomous from all the other procedures of the proposed schema. So much, in fact, as to annul Popper's claim that problems can be regarded as the starting point of the scientific inquiry. For, if this is the case, then the scientist must have a quite well-defined conception (even if it is difficult to articulate it) of what a problem is, how it is to be "recognized" what to "do with it", and what the possible strategies for its solution are (independent of whether they will be successful). How is it, then, possible to justify this following passage from Popper?

> Assume a young scientific meets a problem which he *does not understand*, what can he do? I suggest that even though he *does not understand* it, he can try to solve it, and criticize his solution himself . . . Since he *does not understand* the problem, his solution will be a failure . . . In this way a first step will be made towards pinpointing where the difficulty lies. And this means, precisely, that a first step will be made towards understanding the problem"[19] (emphasis added).

One may not be able to solve a problem, one may not become conscious of the implications of a problem, one may not appreciate how fundamental a problem is, one may not bother too much to see whether a problem is well formulated or not, and so on. But what does it mean not to understand a problem? To answer this question Popper is obliged to tell us explicitly that a problem is "a difficulty, and understanding a problem consists in finding out that there is a difficulty and where the difficulty lies".[20] Thus, the scientist has a vague feeling that there is indeed a difficulty to be solved and explained and then proceeds to do so through the Popperian critical approach. Popper's overall attitude on the "problem" problem is, at the very least, not free of ambiguities, and it surely does not introduce a radical break with the "received" view.

Nickles does propose a satisfactory answer to this question. His "constraint-inclusion" model states that "a problem consists of all the conditions and constraints on the solution plus the demand that the solution (an object satisfying the constraints) be found. The demand arises from disciplinary and programmatic goals and is modulated by the domain of information produced by the discipline and research program".[21]

Since we are interested in the way a physical problem is formulated after the observation of a new phenomenon and in the way it is eventually resolved, the starting point of our inquiry is the understanding of the various kinds of new and unexpected phenomena and the properties they have on account of their being new. Our claim is that among the many formulated problems related to a specific phenomenon, there is one (the right problem) emerging from a paradoxical situation created not simply by a new, but also by an unexpected phenomenon. It is this particular problem that is eventually solved by the research program, and it is only after it is solved that a further test of the solution would be to provide satisfactory solutions for all the other formulable problems. Once we have the right problem, *that* choice is methodologically far more significant than the question of whether the particular problem is badly formulated or not. After the right problem is recognized, sooner or later (and usually sooner) it will become well formulated. The constraints expressed in its formulation determine the dynamic of the process leading to its solution. The form of these constraints had already been "constrained" by the idiosyncracy of the incompatibility between the two contexts which led to the paradoxical situation.

Associated with the constraint-inclusion model of problems, there are a series of difficulties that have to be specifically examined in any systematic analyses of related questions. One such question is the further study of heuristics and heuristic appraisal.[22] The way we treat superconductivity and superfluidity in the course of this book, together with our "reading" of the development of the various theories of superconductivity and superfluidity that we attempt in Part III, clarifies — we would like to hope — some points concerning "heuristics" and the "problem solving potential" of theories and research programs. Here we would like to make some further comments about heuristics, the way they are related to a change of a metaphysical attitude, and their (decisive) role in Lakatos's methodology.

The formulation of a problem through the process we are proposing has another aspect. The complex of constraints inherent in any formulation of a problem helps the researcher decide the type of approaches to be chosen for solving the problem. Thus, the dynamic expressed by the positive heuristic[23] — accepted by most, independently of whether they subscribe to a strong Lakatosian view or not — as a guiding principle of a structured set of constraints together with a strategy to be followed for the solution of a specific problem can be traced to the formulation of the problem itself and the whole process out of which such a formulation of the (right) problem is achieved. This is why, for example, such a way of appraising the "dynamic" is historically more consistent than the proposal to search for the "hard core" of a program together with its protective belt. Apart from the well-known criticisms[24] directed to this point of Lakatos' proposals, it may be that the initially chosen hard core may have been too unconstraining. The danger is, thus, that this "excess" input in the hard core may have been successfully protected. The emphasis on the formulation of the (right) problems leads to a (re-)reading of the "methodology of research programmes" as a methodology in search of a set of procedures which articulate a network of constraints.

In his intervention during the 1965 London Conference, John Watkins remarked that "it seems that a dominant theory may come to be replaced, not because of growing empirical pressure (of which there may be little), but because a new and incompatible theory (inspired perhaps by a different metaphysical outlook) has been freely elaborated. A scientific crisis may have theoretical rather than empirical causes".[25] It is a point whose elaboration can lead to useful clarifications about the heuristic of a theory. High energy physics is a good case to investigate for an elaboration of Watkins's claim.

In dealing with any aspects of high energy physics, the researchers in the field are bound to a Sisyphian methodology. On the one hand, their efforts are aimed at understanding the behaviour of particles which are considered to be elementary, and, on the other, every major development in effect formulates and changes the meaning of the notion of elementarity and provides further evidence to support the claim that the assumption that only certain particles are elementary may have its intrinsic limitations. One way out of this vicious circle is indicated by the fact that the new "building blocks", discovered in the past 20 years, cannot exist freely. The idea of "quark confinement" is not only imposed by the null results of the experimental searches to find them in an unbound state but also because it follows from one of the more successful schemata in high energy physics, that of quantum chromodynamics.[26] The confinement of the new "building blocks" becomes an (inherent) feature of their elementary status and an indispensable aspect of the new and redefined notion of elementarity.

However, the "practical" implications of all these considerations were in the formulation of the (positive) heuristic, which thus acquired an additional regulative aspect: a remarkable constraining mechanism, whereby the multitude of possible theories that the mathematical framework allowed was dramatically reduced, not by the limitations introduced by empirical evidence, but by the limitations which were inherent in the mathematical instantiations of the (new) metaphysical choices specified in the heuristic of a theory. And this is why we tend to believe in Lakatos's suggestion that "it is better to separate the 'hard core' from the more flexible metaphysical principles expressing the positive heuristic".[27] Hence, a regulative function during the development of a programme is implemented not only by the all too obvious regulative principles mentioned by Lakatos[28] (which are, of course, present independent of the details of the different research programmes), but also by the positive heuristic itself. This outstanding feature of the positive heuristic, even though is implied in Lakatos's exposition, is not singled out and its dynamic is not sufficiently expounded.

Lakatos states that the "positive heuristic consists in a partially articulated set of suggestions or hints on how to change, develop the "reputable" variants of the research programme".[29] He then goes on and emphasises that the positive heuristic in a research programme determines the "strategy both for predicting (producing) and digesting"[30] the expected "refutations". The positive heuristic, however, cannot have this function, if it does not also have the regulative aspect we talked about in the form of continuously imposing a series of constraints and limitations on what, on purely mathematical grounds, can be proposed.

> From the beginning it seemed to me to be a wonderful thing that very few quantum field theories are renormalizable. Limitations of this sort are, after all, what we most want; not mathematical methods which can make sense out of an infinite variety of physically irrelevant theories but methods which carry constraints because these constraints, may point the way toward the one true theory.[31]

The change in metaphysics is almost always followed by a new set of constraints. It is this complex of constraining procedures which delimits mathematical physics. In fact, we claim that a programme in mathematical physics turns into a programme in theoretical physics whenever these constraints are explicitly spelled out. It is this kind of change (which does not necessarily hinder mathematical developments) that should be studied further in order to gain an appreciation of the "relative autonomy of theoretical science", to relate it with the whole framework of the methodology of research programmes, and to consider it as an historical fact corroborated by all those historical instances where theoretical work is far ahead of that of the experimenter.

We think that there is much more to relative autonomy of theoretical science than that, and Lakatos seems to be well aware of the shortcomings of his initial exposition: "Let us remember that in the positive heuristic of a powerful programme there is, right at the start, a general outline of how to build the protective belts: this heuristic power generates the autonomy of theoretical science".[32] We do not think, however, that this general outline is so well spelled out, right at the start, as Lakatos suggests, and comprehending the relationship between mathematical and theoretical physics during the development of various programmes becomes a necessary prerequisite for understanding the non-trivial features of the relative autonomy of theoretical science.

We have, lastly, two closely related comments about Laudan's claims concerning the significance of "conceptual problems".[33]

Firstly, there are many cases when the solution of an empirical problem and the emergence of a conceptual problem are causally connected. Taking into consideration such a relationship, however, modifies the "appraisal measure for a theory", as defined by Laudan in the following way: "the overall problem-solving effectiveness of a theory is determined by assessing the number and importance of the empirical problems which the theory solves and deducing therefrom the number and importance of the anomalies and conceptual problems which the theory generates".[34] We shall hold that in addition to this criterion, the problem-solving effectiveness of a theory is augmented by the appearance of a series of conceptual problems which are not usually generated by the theory, but which energe solely because of the successful solution of an empirical problem. Those are conceptual problems which have to be solved *in order to legitimize* the solution of the specific empirical problem: a solution which, at times, may be highly "unorthodox". The empirical problem may have a solution consistent with the experimental data, and yet what may be involved in this solution could be quite "unphysical" and, hence, upsetting for the scientific community. If a scientist attaches much importance to the empirical problem, then what is sought is a way to secure the acceptance of this unorthodox solution by the community. This is usually achieved by solving the conceptual problem generated by the solution of the empirical problem. Such, for example, is the case with the two fluid model in superfluidity which we shall examine analytically in Part II, and with renormalization in quantum electrodynamics and high energy physics.[35] Thus the "appraisal-measure for a theory" mentioned above is complemented by the following: the problem solving effectiveness of a theory is also determined by those conceptual problems whose solutions give further credence to the "unorthodox" solutions of empirical problems (and increase the solutions' acceptability by the scientific community). It is such an additional criterion which provides a "rationale" for the (correct) claim that "it can be rational to pursue investigation of a theory even if it is irrational to accept it".[36]

The second comment is this. This "paradoxical situation" we have defined above acquires a quite idiosyncratic function: it "transforms" an empirical problem into a conceptual problem. What we have called the "right" problem is really a conceptual problem, since it has been the result of a paradoxical situation which, in turn, has emerged after the observation of a unexpected phenomenon and after the attempts to formulate it as an empirical problem in the descriptive language of the dominant theory. Furthermore, the emergence of a paradoxical situation and the ensuing (right) physical problem is really both a conceptual and an empirical problem. It is the latter since one has to explain the observed unexpected phenomenon, and it is the former since it is the implications of the description of the specific unexpected phenomenon that revealed the incompatibility with the dominant theory, and, thus, created the paradoxical situation. It follows that the explanation of an unexpected phenomenon is a solution both to an empirical and a conceptual problem. The Meissner effect, the "viscocity paradox" of liquid helium, and the fact that helium remains liquid down to absolute zero are all cases in point and will be presented in a detailed manner in Part II.

1.4. Concepts out of context(s): How problems are solved

The formulation of a new phenomenon into a well-posed physical problem follows those instances wherein a paradoxical situation is created by the initial attempts to describe the new phenomenon. The paradoxical situation does not arise because the observed phenomenon has not been predicted by any theory. This situation also does not arise because the phenomenon defies description in terms of concepts borrowed from the "accepted" theoretical framework or from attempts to find an explanatory schema with the consecutive approximations that these concepts "allow". That is, the paradoxical situation is not between the phenomenon and a framework which cannot initially provide a consistent explanation, but rather it lies between the results of the extrapolations allowed by the concepts used for the description of a new phenomenon and the theoretical framework which provided the concepts for this description. Such a paradoxical situation is not a logical inconsistency. It is best described as an incompatibility between two implied "ontologies": one implied by the accepted theoretical framework and the other by the description chosen for the new phenomenon.

Solving a problem means creating those conditions in which there is no longer a paradoxical situation. The two contexts referred to above are not irreconcilable after the proposal of a successful solution. Thus, the solution of a problem may involve the introduction of new concepts within the existing theory; it may involve the reinterpretation of various aspects of the theory; it may involve the proposal of new mechanisms; and it may also involve incorporating violations of "holy" rules into the newly proposed theoretical framework. In other words, solving a problem arising out of a paradoxical situation created by the description of a new phenomenon involves procedures which eventually transform "that which is unexpected" into "that which should be expected". The changes incurred by the solution to the problem expand the range of what we consider as expected behaviour. These changes thus "legitimize" what was unexpected behaviour before the solution as behaviour which indeed has its place within the newly formed framework. Eventually, the new and unexpected phenomena will be assessed against this newly formed framework which now becomes the "standard of normalcy" and so on.

We propose a process we shall call "concepts out of context(s)" as a process that appears to be followed in solving the specific class of problems we are examining. By such a process, we refer to a *generic* characteristic of the overall approach that is adopted for the solution of these problems, rather than a specific algorithm to be prescribed for all such cases. It is through this process of "concepts out of context(s)" that the incompatibility between two contexts will eventually disappear. A new theoretical framework will emerge, and it will be possible to claim that we have a satisfactory explanation of the new and unexpected phenomenon.

Let us be more specific:

"Out of" has a double meaning, and both meanings are at work during the process of solving a specific problem. In the first case by "concepts out of context(s)" we shall refer to that process by which concepts are derived from the initial framework, a process which takes the concepts from "within" the theoretical framework and situates them "outside" of it. This procedure, when needed, will be referred to as (out of)$_1$. In the second case, by "concepts out of context(s)", we shall refer to that state of affairs where the concepts, having already been derived (out of)$_1$ the initial theoretical framework (TF_i) have also a certain degree of independence, a certain degree of autonomy with respect to the framework (out of)$_1$ which they were derived. This aspect of the concepts we shall call (out of)$_2$.

After one decides on the strategy to be followed to resolve the paradoxical situation, one chooses a set of concepts (and mechanisms) as the more likely to start addressing the problem. If all these concepts are to be referred as c_i, then $\phi(c_i)$ will partly denote the description of the new phenomenon in the language of the specific theoretical framework we choose, partly its formulation as a physical problem, and partly the attempted (tentative) explanation.

These concepts are not derived deductively from TF_i. They do not, in other words, belong to the conceptual "hierarchical structure" of TF_i but are "motivated" by TF_i. These concepts have an element of "arbitrariness" which is not even *a posteriori* reconctructible and cannot be explained rationally. This is why the concepts devised to formulate the new and unexpected phenomena into a physical problem are concepts *out of* the initial context, and, more specifically, out of TF_i.

The meaning of concepts is not only due to their definitions or to the fact that some are derivable from others. That is, their meaning is not merely determined from their position in the conceptual "hierarchical structure" of a particular theory. The role of concepts during a specific stage of the problem solving process is also a determining factor. A concept thus acquires meaning due to its (de-)contextualization: concepts are first "taken away" from their original context, and their ensuing relative autonomy from, say, TF_i eventually allows the formation of a new context. This autonomy has been achieved due to the excess meaning the concepts acquired from their use in attempting to explain the unexpected phenomenon. When a particular concept is chosen for such a process, there is an additional meaning to this concept which has been acquired because it finds itself in a new context different from the one (out of)$_1$ which it was derived. Therefore, $\phi(c_i)$ is (out of)$_1$ TF_i, but also is in a state which is (out of)$_2$ with respect to TF_i.

We are therefore interested both in the meaning concepts have because of their place in the conceptual hierarchical structure of a theory and also in the meaning they acquire due to their function in the specific process of attempting to find an explanation to the new and unexpected phenomenon. We are, in other words, interested in the meaning concepts acquire due to their place in a *context*. This is, in a way, quite similar to Nersessian's (1984) claim that a conception of meaning adequate for scientific theories could be achieved by studying the actual scientific practices concerning meaning in addition to the questions related to the study of language.

The conceptual "hierarchical structure" alludes to the fundamental concepts and relations and to the process of "decoding" by predominantly deductive means the wealth of information "hidden" in these relations. On the other hand, the conceptual context, without undermining what is implied by the conceptual hierarchy and by complementing it, introduces a "bootstrapping" procedure for the concepts to acquire meaning. This meaning is not only determined due to a well formed structure with clearly and unambiguously defined relations among the various concepts, but also because of the fact that the way they formed the new context creates a non-systemic aspect to the overall conceptual structure such as to provide its own meaning.

It is the contextual character of their acquired meaning which gives these concepts an excess (meaning) content (something which, in this sense, does not exist for concepts whose meaning is only due to the conceptual "hierarchical structure" of the theoretical framework) which cannot be accounted for through the procedures used to derive concepts from more basic ones and form a hiererchy. It is this excess both of content and of arbitrariness which allows (out of)$_1$ → (out of)$_2$.

We shall not discuss the sources of this excess meaning, even though their study presents a tremendous interest: these sources are to be sought in the metaphysical beliefs of the (particular) scientist(s), in their ideological assumptions, in their prejudices, in their methodological stance, etc.[37]

Up to now we have:

TF_i

↘

(out of)$_1$ *and* $\phi(c_i)$ is (out of)$_2$ with respect to TF_i

↘

$\phi(c_i)$

Having, thus, $\phi(c_i)$ *out of* the TF_i, one goes back to TF_i attempting to "pack" and to fit $\phi(c_i)$ into the conceptual structure of TF_i. Such "packing" is attempted through a series of (re)interpretations, (new) assumptions, (new) mathematical developments, (new) foundational principles, (new) physical mechanisms, etc.

If now $\phi(c_i)$ could be fully accommodated into TF_i, then this means that the TF_i can indeed provide an explanation for the new phenomenon. A more common situation is, after such an attempt, to change TF_i into TF'_i and then have $\phi(c'_i)$ which is *out of* TF'_i, and so on.

Going back to TF_i to examine the possibility of whether the $\phi(c_i)$ can be accommodated in it is but a "natural" undertaking of the scientific community. It may indeed be ascribed to a conservatism of the community which does not generally want to dispense with basically successful explanatory schemata. Nevertheless such a procedure is obviously more practical and, historically at least, more effective than attempting to start "all over again".

We are not, however, interested in those aspects as such, but in what it is in our approach that allows this "going back". What is it intrinsically that legitimizes, on methodological grounds, going back and attempting to coalesce the $\phi(c_i)$ into the TF_i? One follows this procedure because of the excess meaning of the c_i when the particular c_i is used in $\phi(c_i)$ even thought the c_i was originally in TF_i. It is then justifiable to go back to TF_i exactly because $\phi(c_i)$ is in a state of (out of)$_2$ with respect to TF_i, for the very reason that $\phi(c_i)$ has been derived by the process of (out of)$_1$ from TF_i. $\phi(c_i)$ is out of TF_i *because* the particular concept c_i is *out of* the context of the specific theory, and *inversely* it becomes possible to explain the ϕ and/or postdict or predict the ϕ.

We know we have reached the new theory when there is a $\phi(c_i^j)$ that is derived out of TF_i^j which has practically no excess meaning, and it can be "fitted back" into TF_i^j almost trivially. In other words, it is the moment when the newly formed framework $[TF_i^j \equiv TF_n]$ is able to provide an explanation for the (originally) unexpected phenomenon, which is no longer surprising within the new theoretical framework.

The change of TF_i into TF_i' and of TF_i' into TF_i'' etc. and the eventual "emergence" of TF_n through this continuous interaction of the TF_i's with the $\phi(c_i)$'s become possible because of the *intermediate intermediaries*. These are concepts which "bridge the gap" between a TF_i^j and $\phi(c_i^j)$. In nearly every case, the intermediate intermediaries express a particular manifestation of the c_i's which during a particular period in the development of a theory are called to play a predominantly methodological role. The progressive change of TF_i into TF_n is achieved by the repeated attempts to accommodate the concepts used for the description of the new and unexpected phenomenon ϕ into TF_i's through a series of procedures allowed by the intermediate intermediaries. Hence, the change of TF_i is accompanied by a subtle metamorphosis of the various concepts.

The intermediate intermediaries are concepts of a *predominantly procedural character*, and they are "responsible" for the process by which one attempts to embody and coalesce the newly devised concepts into the theoretical framework (out of)$_1$ which they were originally "formed". The ensuing changes in the initial theoretical framework (TF_i) "lead" to ("cause" . . .) progressive transformations of the $\phi(c_i)$ into $\phi(c_i')$, of $\phi(c_i')$ into $\phi(c_i'')$ etc.

When TF_n is finally formed, the concepts initially used for the description of the unexpected phenomenon, and then for its formulation into a physical problem, are either part of the basic concepts of TF_n or derived by them. These concepts, c_n, become independent of the original context; they are beyond the range of the implications of the original context; they have already formed their own context; and they now belong to the conceptual "hierarchical structure" of a new theoretical framework.

In our terminology the constraints are manifested during the ensuing process of "concepts out of context(s)" and they are "articulated" through the regulative function of this always present yet never explicitly defined relationship between $(\text{out of})_1$ and $(\text{out of})_2$. How much is "borrowed" from a *TF*? What concepts are to be chosen and "streched" from *TF*? What is the excess meaning of a particular $\phi(c)$? how is the meaning autonomy of $\phi(c)$ to be exploited? How is one to "re-attack" the TF's in order to "fit" the enigmatically acquired excess meaning? These are some aspects of the regulative role of the double meaning of "out of". The totality of all these aspects "makes up" the heuristic strength of each specific inquiry. Lakatos' "positive heuristic" and Kuhn's tension between tradition and innovation find their analogue in this relationship between $(\text{out of})_1$ and $(\text{out of})_2$. (See Figure 1.)

1.5. Theoretical and experimental tests

For a systematic investigation such as the one we are attempting of the methodological problems related to the formulation of observed new phenomena into physical problems and the ensuing procedures for the solution of these problems, it is necessary to discuss certain features of experimentation in its more general sense. In the following we shall discuss some characteristics of the relationship between theory and experiments and the kinds of tests involved in examining theories and probing theoretical frameworks.

The specific development of the different theories proposed to explain the phenomena occurring in low temperature physics reveals some interesting aspects concerning the relationship between theory and experiment. This relationship has been found to transcend the problematic about the nearly exclusive role of experiments in either verifying or falsifying theoretical predictions. The study of various research programmes has shown in no uncertain terms that the totality of the experimental procedures attains an unique historical character. The realm of what is considered to be physically real is not confined exclusively to given experimental results. There are instances during the formulation of a theory when the correctness of existing experimental results is disputed because of what is considered to be the heuristic strength of the proposed schema.

Despite the extensive reference to the relationship between theory and experiment in many works on philosophy of science, relatively little attention has been paid to the systematic study of experiments and experimental procedures, not as specific case studies, but rather, as an attempt to investigate the variety of methodological problems associated with this particular activity of the scientific community.[38] Prior to such a study, however, one is obliged to think about the possible divisions of the tests to which theories are subjected into different categories and to recognize the differentiating aspects of each category.

Description of the observed unexpected Phenomenon (ϕ)	Paradoxical Situation	Physical Problem	Process of Solution	Solution
$D(\phi)$	Incompatibility between $J(TF)$ and $J(D(\phi))$	$\Phi(c)$	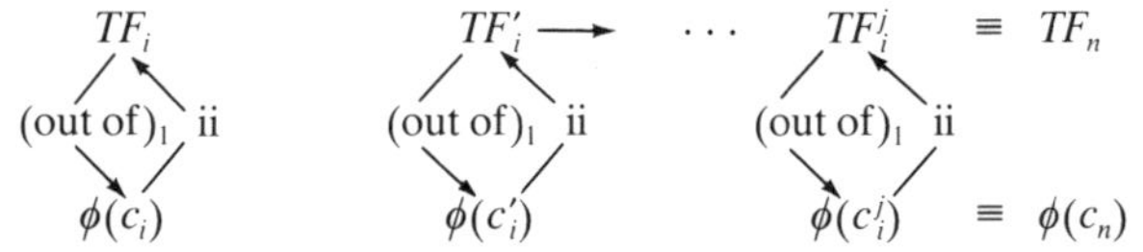 TF_i ⇄ (out of)$_1$, ii ⇄ $\phi(c_i)$; $TF'_i \longrightarrow$ … $TF^j_i \equiv TF_n$ ⇄ (out of)$_1$, ii ⇄ $\phi(c'_i)$ … $\phi(c^j_i) \equiv \phi(c_n)$	$c(\phi)$

$D(\phi)$: The observed unexpected phenomenon expressed by the descriptive language at the dominant theoretical framework TF.

$J(TF)$: The implication of the dominant theoretical framework which does not only include the deductions from the best available theory, but also the influences of other (metaphysical, theoretical, epistemological, etc) considerations in forming these implications.

$J(D(\phi))$: The implications (and connotations) of the description of the newly observed unexpected phenomenon much in the same spirit as in $J(TF)$ since one uses the descriptive language "of" *TF*.

$\Phi(c)$: The formulation of the problem by making use of specific concepts of the dominant theoretical framework.

$\phi(c^j_i)$ is always (out of)$_1$ TF^j_i *and simultaneously* is (out of)$_2$ with respect to TF^j_i. When it is seen that $\phi(c^j_i)$ is *not* (out of)$_2$ with respect to TF^j_i, then that means that we have reached the solution, $c(\phi)$, of the problem. The change of TF_i into TF'_i and of TF'_i into TF''_i etc. and the eventual "emergence" of TF_n through this continuous interaction of the TF^j_i's with the $\phi(c'_i)$'s becomes possible because of the intermediate intermediaries, (ii).

Fig. 1.

The notion of experimentation and, more specifically, that of testing cannot be confined to proposing, planning, and performing the experimental tests. If, however, we should — as we must — consider "experimentation" as the totality of those procedures which make explicit the implications of the set of possible constraints imposed on the theory and the further limitations expressed by them, then the notion of experimentation includes much more than the implications of the measuring procedures. Thus, experimentation is neither merely the process in which "one tests an hypothesis under controlled conditions" nor it is merely a process enabling the scientist to "ask questions of Nature and receive answers". Experimentation is a generalized process of probing both into the proposed theoretical entities and their assumed relationships, and bringing forth the features and structure of the constraints inherent in these assumed relationships among the theoretical entities. *Hence, experimentation becomes the process of investigating the limitations of a theory.*

An almost universally accepted criterion for a theory to be considered successful is that it at least explains existing data in a satisfactory manner and that it predicts certain new facts. Such a theory is subjected to a series of both theoretical and experimental tests during each stage of its development. An additional criterion for the success of a theory is that it "reacts positively" to those tests.

These tests can be grouped in the following general categories:

i. Grouped as a set of *theoretical conditions* which are imposed on the theory, one tries to extract the maximum information from the theory itself. Such conditions, in many cases, provide heuristic devises which facilitate the formulation of the basic equations. The set of these conditions and the way one chooses to impose them draw out the long-term strategy for further investigations related to the proposed theory.
ii. Grouped as a set of *generalized theoretical approaches*, one can decide what kinds of experimental tests for the specific theory are the "most appropriate". This is a process of translating the newly proposed notions into testable statements and of convincing the community that the newly introduced concepts are, in fact, testable. This process should not be confused with the proposal of specific experiments to be performed for testing the theory.
iii. Grouped as the kinds of the actual *experimental tests.*

Let us now discuss each category separately.

i. Theoretical conditions

No theory can be constructed without being subjected to a set of theoretical tests during specific stages in its development. These are the following:

1. Tests of consistency and completeness

These conditions have to be satisfied in order to avoid contradiction in the theory and to start deriving a series of statements which would follow from the fact that we have a contradictory theory, than having these statements be an indication of the strength of the predictive power of the theory. These are tests that are routinely performed in the axiomatic construction of various theories with sound mathematical foundations.

2. The "low" and/or "high" limits

An absolutely necessary condition for convincing the scientific community of the correctness of a newly proposed theory is being able to successfully derive from the new theory either the theory which was previously in use or — in the event that there was no such overall theory — the salient aspects of the models being used.

The limiting procedures involve not only mathematical techniques, since it is important to understand during the limiting process what aspects of the theory need a reinterpretation and what aspects should remain "unchanged". The limiting case should not only display an isomorphism with the previous theory or model; it should also preserve the same physical interpretation with, whenever this is possible, no new *ad hoc* or additional assumptions being introduced after the limiting process has been completed in order to give the "accepted" interpretation. The limiting procedures, however, have an additional function in those cases where they provide the clues for new and "unusual" physics and spell out the difficulties involved in the proposed theory of whether the new physics suggested by those limiting cases is readily interpretable.

3. Initial/Boundary values

The boundary values are conditions accompanying a differential equation in the solution of physical problems. In mathematical problems arising from physical situations there are two considerations involved in finding a solution: (a) the solution and its derivatives must satisfy a differential equation, which describes how the quantity behaves within the region; (b) the solution and its derivatives must satisfy other auxiliary conditions either describing the influence from outside the region (boundary values) or giving information about the solution at a specified time (initial values), representing a compressed history of the system as it affects its future behaviour.

The relationship between physics and mathematics is important here, because it is not always possible for a solution to a differential equation to satisfy arbitrarily chosen conditions; but if the problem represents an actual physical situation, it is usually possible to prove that a solution is possible even if it cannot be explicitly found.

4. Questions allowed to be formulated by the theory

What emerges as an interesting feature of competing theories or programmes is that they allow us to pose a series of new questions thereby redetermining the strategy of problem solving.

The ability to pose more questions is an additional expression of the "increased" empirical content of the specific theory and, in that respect, the actual formulation of the questions can indicate directions of research and/or growth of the programme. What, however, is quite interesting is that such a criterion can be used to evaluate programmes developed as alternative modes of explanation to the dominant theory with predictions similar to those of the dominant theory. Yet one cannot dismiss these programmes which duplicate known facts if, within their contexts, one is allowed to pose a series of questions which were not allowed within the context of the dominant theory. Furthermore, these programmes should not be shunned even in situations where the solution to the problem(s) implied by the(se) question(s) may not be different from the one provided by the dominant programme, exactly because being allowed to pose new questions is an indication of a (peculiar, no doubt) increase in the empirical content of the propositions (or models) of a programme.

Lakatos, in a comment about the "creative shift" that may boost a degenerating programme, remarks that

> On should not forget that two specific theories, while being mathematically (and observationally) equivalent, may still be embedded into different rival research programmes and the power of the positive heuristic of these programmes may well be different. *This point has been overlooked by proposers of such equivalence proofs*[39] (emphasis added).

The assertion that, in considering rival programmes, one has to take into account the kinds of questions which are now posable introduces an indispensable new criterion for seeking and establishing "equivalence proofs". Such a criterion, however, exactly because it eases the observability criterion, makes the establishment of equivalence more difficult since there are quite serious epistemological, methodological, and sociopsychological problems associated with the notion of proving the equivalence of alternative programmes. It can only be considered as a technique which establishes trivial correspondences (often with very sophisticated techiques) and as a process which "takes away" from the programmes the richness of their individuality.

This existence of many different theories which provide equally satisfactory explanations for the phenomena, and with "comparable" empirical contents, should not prompt us to think of them as equivalent modes of explanation and seek a formal way to prove it. Instead we should attempt to construct metatheoretical frameworks to "accommodate" the different programs and, by "accommodating" them, to comprehend not their equivalence but the deeper reasons for the complementarity of the explanatory possibilities they provide. For this purpose, the study of the spectrum of the questions which can be formulated becomes a crucial undertaking.

5. Questions allowed to be formulated by the theoretical framework

There are periods when the scientific community is confronted with a series of competing theories while the existing experimental data do not exclude any theory. It may then become necessary to construct a framework in order "to compare the theories on an equal footing". The construction of such a framework involves quite intricate methodological problems, and one can use the framework to clarify certain conceptual problems which are found in each theory separately. Such is the situation which started with Dicke's reinterpretation of Eötvos's experiment and culminated in the construction of the "parametrized post-Newtonian" framework which was a formalism to compare metric theories of gravity with each other and with experiments. The uniqueness of such a programme is that it is not composed of a succession of theories that tackle a series of anomalies. It consists of a cohesion of theoretical frameworks to "accommodate" the various alternatives to the General Theory of Relativity. Hence, the prediction of novel facts[40] cannot (by definition, more or less) be sought for, either in the explanation of an anomaly or in the prediction of an hitherto unknown phenomenon. The novel facts in such a programme acquire the status of methodological novelties and are expressed as realizable possibilities to pose new questions, to provide experimental tests for fundamental principles, and to proceed to the "foundations of a theory of gravitational theories". This "metatheoretical" schema is the outcome of the research programme for the construction of theoretical frameworks for experimental relativity.

6. Invariance principles

Discovering the transformations which leave invariant the equations of a theory means at the same time that the specific equations have an underlying symmetry. What is of further interest from our point of view is that such an invariance implies a conserved physical quantity. Hence, an approximate or even a badly violated conservation law provides us with equally important information as when such a law is perfectly obeyed. Again, an important problem after the discovery of these transformations is to give a physical interpretation of the ensuing conserved quantities and to provide a physical interpretation for the symmetry breaking mechanisms, since most symmetries are only approximatly valid.

These are cases for which the proposed symmetries imply too "perfect" physical systems, something that we know from the beginning could not be true. Nevertheless, these symmetries are proposed, because they resolve specific theoretical problems in the construction of the theory, and, immediately afterwards, one searches for the breaking mechanisms, for the physical interpretation of these mechanisms, and for their experimental manifestation.

Another aspect of the transformation requirements is to raise a particular invariance property to the status of a principle and to demand that the derived equations obey the specific transformation property without providing a physical interpretation of the specific mathematical transformation.

ii. Theoretical approaches to experimental proposals

The development of a new theory is nearly always followed by a specific proposal about the ways in which it is possible to test the more salient features of the theory. Before the actual tests are performed there is a series of different methodological approaches to decide which specific experiments are the most appropriate for the proposed theory. We will not be referring here to the specific experiments which happen to be more sensitive to the proposed theories, or to the intricate technical problems which have to be overcome, but to those aspects of any proposed theoretical schema which provides an overall orientation as to the kinds of quantities to be measured. There are four such theoretical approaches.

1. Approximation

This approach is used when it is realised that the connotations implied by the idealizations involved in the proposed theory will not have an adverse effect on measurements involving the real entities. In such cases one proposes experiments whereby one can neglect the effect of all those factors which could undermine the assumed idealization. Usually the practical result of such an approach is the neglect of factors whose overall contribution is "too weak".[41]

2. Perturbation

This is, in effect, the opposite of the previous approach. For the idealisations involved in every theory, there is a well-defined method for introducing "elements of reality" in order to make the theory conform with those aspects which have already been measured experimentally or which comprise indispensable facets of the accepted ontology.

For the proposed experiments one has to take into consideration a totality of factors in which "small values" cannot be regarded as the sole factor in determining whether these factors should be neglected or not, since either their cummulative effect or their qualitatively different effect provides a quite different picture than the one expected from the idealised theory.

3. Concept innovation

One of the more interesting types of theories is the one whose predictions are new phenomena rather than the prediction of values of quantities which belong to known phenomena, but which have not been measured thus far. In such situations one is obliged to devise an unusually complex "transformation language" in order to specify the way the proposed new physics is accommodated either in the existing observation language or, if not, in its proposed extensions. Hence, apart from the new concepts being introduced for the development of the theory itself, there are additional concepts that have to supplement the existing observation language.[42]

4. Compactification

There is a series of theories in which the attempted idealization is not only counterintuitive but also highly "unphysical". Compactification refers to the strategies for testing such theories and we shall discuss them briefly. The recent situation in high energy physics is again quite instructive. After a set of experiments which provided ample corroborative evidence for the theories being tested, all subsequent experimental results were either predicted or explained in a relatively straightforward extension of the theories. In the meantime, however, there was the development of alternative rival schemata and models, and the new phenomena predicted by these schemata involved experimental set-ups that, even under the most favourable conditions, one could not envisage their realization within the next quarter of a century.[43]

At present there seem to be no outstanding experimental anomalies. Existing data seem to be explainable by the various proposed schemata. But the possibility of subjecting the rival schemata to imminent tests for the truly new and differentiating phenomena they predict is removed to a future nobody dares to predict. How, then, did the community of physicists meet this unique challenge? "In this unfortunate circumstance, when theorists are not provided with new experimental clues and paradoxes, they are forced to adopt new strategies".[44]

Researchers were obliged to formulate a series of strategies in order to guarantee that the proposed schemata would be imminently testable and that the assessment of their validity would not be removed to an unreachable future. These strategies necessarily had an influence on the way one would construct the "protective belt" of the programs, and the way one went about formulating an overall observation language.

These strategies developed along two lines: the first involved the way a mathematical construct was to be reduced to make "contact with reality"; where the reduction does not involve a series of approximations. Nor is the reduction the opposite process, whereby the idealisation involved in the mathematics should be perturbed by the characteristics of the real world. What we have is the following situation: a series of problems could be resolved if one makes a set of "unusual" assumptions some of which we know are not valid in the real world, defined as the physics we do in four dimensions at energies less than 1 TeV. These assumptions give rise to structures far removed from the ontology of four dimensional physics. One kind of strategy, then, involves procedures devised to give (back) to the various derived mathematical structures a physical status which has been undermined through the highly unphysical set of assumptions. What is remarkable is that such mechanisms have recently been proposed and seem to be quite promising for the string theories.

The reformulation of the whole strategy for the experimental testing of these theories resulted in the redefinition of what we usually accept as being the role of the limiting cases. Instead of devising approximation procedures for low (energy, mass or whatever) limits, an attempt has been made to single out some of the already known phenomena to be predicted in the limiting cases as being "crucial" despite the fact that the crucial new phenomena predicted by the theory itself are for unimaginably high energy regions. One has to be careful here, in order to avoid being confused by the following situation. When we routinely use limiting approaches among the data we want to "reproduce", there are data that our past experience has convinced us are "more crucial" than the rest, and it is usually that specific class of phenomena which have a precedence in our testing. Those phenomena are preferred because of our past experience, and if they do not result as predictions of the theory after the application of the approximation procedures, then they cannot be considered as reflecting on the effectiveness of our reduction strategy.

A second strategy concerns the decisions involved in the more subtle questions concerning the interpretative context of the various programmes. Every successful theory predicts a series of new phenomena and involves a number of more or less *ad hoc* parameters. Some of them are determined solely by the predicted new phenomena. An additional strength of a theory is expressed by a situation whereby existing data can be postdicted after the new parameters are determined from the new phenomena. There are, of course, cases in which one has the opposite situation: the new parameters are not determined uniquely by the new phenomena and, therefore, their determination from the existing data leads also to (more or less exact) quantitative predictions concerning the new phenomena. There is also a third case which is methodologically acceptable, even if a little dubious on ethical grounds. This is to find the values of the new parameters by demanding that the theory should be able to explain existing data and, thus, predict the quantitative aspects of the new phenomena as well. If the possibility of detecting the new phenomena is removed into the distant future, and the total number of parameters involved is not small, and the errors involved in the existing data are of the usual kind, then this procedure is one with which you cannot go wrong! But physicists, on the whole, are an honest lot and quite demanding in searching for a convincing methodology that makes "contact with reality."

iii. The experimental tests

In discussing the types of experimental tests and the kinds of experimental results, it is not possible to investigate them without taking into consideration both the state of the theoretical developments during the specific period in which the experiment was actually performed and the dominant problematique concerning conceptual questions as well as the implications of the interpretation of the experimental results within the existing theoretical framework. Each kind of experiment may make use of more than one experimental technique and design without this fact altering the category to which the experiment belongs.

Let us now discuss each type.

1. The first type are experiments performed after the proposal of a theory and are aimed at the testing of specific predictions of the theory, where we know before actually doing the experiment, that the deviations from the predicted values would be of a statistical nature. These tests make up the most common kind of experiments and are usually the result of the "approximation" approach.
2. The second type of experiment is similar to the first in that these experiments are performed after a theory has been proposed to test specific predictions, but we know before actually doing the experiment that the deviations from the predicted values are the result of "other laws" the consequences of which cannot be isolated. These "laws" are either known to us or we may be in a position to say something qualitative about them. Investigating the deviations from the predicted values, after isolating the statistical fluctuations in the measured values, is a procedure commonly used either to improve a theory by eventually discovering the structure of the effects causing these deviations or to use the persistently unexplainable deviations as a criterion for rejecting the theory.
3. Another category of such experiments whose overall effect leads to a redetermination of the theoretical context of the relevant theory are the so-called null experiments. These are extremely accurate experiments, and they test what turn out to be implicitly accepted assumptions. These experiments are of fundamental importance in attempting to "improve" a theory. It is precisely because of the acquired "foundational status" of these assumptions that an investigation of the implications of null experiments may lead to a reinterpretation of the theoretical context used and, hence, may introduce an alternative

ontological background. Such a programme was initiated in gravitational physics by Dicke's remarkable reinterpretation of the Eötvös experiment.

A subcategory of the null experiments comprises the measurements of those quantities which the theory predicts to be non-zero and which turn out to be almost exactly zero. Unless there is an acceptable hypothesis to protect the fundamental assumptions from such a drastic anomaly, these results are an indication that the theory is in deep trouble!

Finally, one can include in this category the experiments performed to test the conservation laws, but these experiments, as important as they may be, do not have the basic characteristics of the null experiments and should be considered as extreme cases of categories 1 and 2.

4. The fourth category consists of experimental measurements of different aspects of a new phenomenon where the results — as is discovered later — are either wrong measurements or arise out of "imperfect" experimental setups. The mistakes are discovered either after repeating the experimental measurements or after the proposal of a new theory or model whose predictions compel us to redo the experiments. These kinds of experiments usually precede the formulation of a theory and what is remarkable is that in most cases such experimental results — even though they do not seem to provide corroborating evidence for the proposed theory — are not considered to be factors hindering the development of a theory. Lakatos, for example, emphasizes this situation, even if he exaggerates the dynamic of the positive heuristic.
5. Experiments in category 5 produce unexpected and unexplainable results that seem to be due to inborn factors of the measuring mechanism. These kinds of results are usually an indication that some fundamental principle may not be so fundamental. A characteristic example of such a situation is an experiment performed in 1928. Cox and his collaborators wanted to carry out an experiment with electrons in analogy with the optical experiments which established the phase waves of de Broglie and Schrödinger. In this experiment, they found a deviation from the expected value of a certain parameter, indicating a preference in the direction of the outgoing electrons. Despite some technical points, involving an error in the reported sign of the asymmetry, there can be no doubt that these experiments furnished the first evidence for the non-conservation of parity. This conclusion was never mentioned by the authors or anyone else at that time, and the asymmetry was thought to be due to factors concerning the apparatus itself.
6. This category consists of experiments in which the non-zero value of specific parameters overthrows the observation language used, since, usually, the value zero is an indication that a particular principle is obeyed exactly. In gravitational physics, for example, the specific physical meaning of the coef-

ficients of the "supermetric" of the parametrized post-Newtonian formulation is the result of further approximations and assumptions introduced on the post-Newtonian framework. By relaxing, substituting, adding, or rejecting some of these approximations and assumptions, one could have a different spectrum of physical attributes for the parameters and, hence, a different set of statements which make up the observation language. There is, for example, one specific parameter in the ppN formalism which plays a dual role: its zero value is an indication that both a conservation law is obeyed exactly and that there can be no preferred frame. What would, for example, be the implications of a non-zero value of such a parameter?

Nearly all the facets of the new phenomena we talked about as well as the various theoretical and experimental tests can be traced in the development of low temperature physics, with which we deal in the following three chapters.

PART II

The What

You may find agreement whith theory: then you have carried out a measurement. But if you are lucky you will find disagreement. Then you have done an experiment.

E. Fermi

Starting from the idea, that the phase transition into He II might be interpreted

<table>
<tr><td rowspan="3">as</td><td>the realisation</td><td rowspan="2">of</td><td rowspan="3">the condensation phenomenon of the Bose–Einstein statistics . . .</td></tr>
<tr><td>a manifestation</td></tr>
<tr><td>due to</td><td></td></tr>
</table>

F. London (from a letter to his brother Heinz, 28 July 1939).

CHAPTER 2

Early research at Leiden and some of its methodological implications

2.1. Preliminaries

It was in the Physical Laboratory of the University of Leiden that helium was first liquefied in 1908, that superconductivity was first observed in 1911, that helium was first solidified in 1926, and many properties of what eventually came to be known as superfluid helium were first discovered. Many other developments — of a less "dramatic" nature, but as important to the development of low temperature physics — took place in Leiden which, between 1908 and 1923, was the only laboratory in the world[1] where liquid helium could be produced and used to make physical measurements. These achievements were the result of a long-term strategy first set many years before 1908 by the Laboratory's Director, the person who was known among the members of the Institut Internationale du Froid as "le gentleman du zero absolu" or as "Mr. Freezer" for some Dutch cartoonists, the person appointed to the first chair of experimental physics in the Netherlands, Heike Kamerlingh Onnes (1853—1926).[2]

The purpose of this chapter is to analyse the various elements in the strategy followed by Kamerlingh Onnes in his researches, and to proceed to a discussion of those methodological trends which are implicitly expressed in these investigations. There will be an attempt to classify and study the various directions of research in Leiden from the time of Kamerlingh Onnes' appointment — after succeeding Ryke — as professor of experimental physics at Leiden in 1882 to the time of his retirement in 1922, with reference to all the subsequent papers bearing his name which appeared until the time of his death in 1926.

To appreciate Kamerlingh Onnes' researches fully one has constantly to bear in mind that experimental ingenuity went hand in hand with sophisticated theoretical pursuits. Kamerlingh Onnes, throughout his researches, insisted on the principle of "knowledge through measurement" which he so eloquently expressed in his inaugural speech at Leiden, while at the same time he was fully aware of the limitations of such an approach unless one had a firm command of the theoretical developments and an ability to contribute to those developments.

In the preface to his thesis, he quotes Helmholz when the latter delivered the memorial lecture on Gustav Magnus.

> It seems to me that nowadays the conviction gains ground that in the present advanced stage of scientific investigation only that man can experiment with success who has a wide knowledge of theory and knows how to apply it; on the other hand, only that man can theorize with success who has a great experience in practical laboratory work.[3]

Kamerlingh Onnes, throughout his researches, followed this dictum closely, and his theoretical contributions were considerable: what should be emphasized, however, is not whether his theoretical contributions were insightful or lasting, but rather the fact that he considered this activity to be an integral part in the planning of his investigations. The planning of the experiments was not guided only by the needs of previous experiments or by theories and propositions put forward by others, but also by hypotheses proposed by Onnes himself. Nearly none of these hypotheses survived, and his whole "behaviour" shows how little prejudiced he was towards his own hypotheses. Even when there was sufficient data to claim their "confirmation", he was always on the *qui vive* about the appropriateness of the apparatus or the experimental set up, and generally sought weaknesses in the measuring process or oversimplification of the proposed hypotheses.

It is possible to divide Kamerlingh Onnes' researches into the following categories: researches on the equation of state, electrical researches, magnetic researches, properties of liquid helium and, one can add to these main directions, the study of the — electrical and magnetic — properties of gases and metals at liquid hydrogen and helium temperatures. Another research direction involves instrumentation and the constant improvement of thermometric methods.[4] There is, obviously, a strong interrelationship among all these directions, yet there is also a relative independence of the researches pursued within each one of these categories.

In this section we shall discuss the thermodynamic and magnetic researches, and we shall include the electrical researches and the researches on the properties of liquid helium in the same general and systematic treatment of superconductivity and superfluidity. Before proceeding to a relatively detailed presentation of these research programmes in order to understand Kamerlingh Onnes' overall strategy, his motivations for drawing up such a strategy, and the importance of his theoretical contributions together with that of the less fantasmagoric aspects of his investigations, however, let us present the main conclusions from our "reading" of his researches.

i. To comprehend the underlying basic motivating aspect of Kamerlingh Onnes' researches which made him unswervingly follow his programme, one has to examine the work Onnes did between receiving his doctorate in 1879 and his appointment in Leiden in 1882. Kamerlingh Onnes, right after his appointment in Leiden, draws up a program for the change of phase of hydrogen and, after Dewar liquefies hydrogen in 1898, he aims his researches at liquefying and solidifying helium: his main motivation was that such an achievement would be decisive corroborating evidence for the law of corresponding states first put forth by van der Waals in 1880[5] and for a theorem also proved independently by himself in 1881.[6] This fact is either not known or, when known, its significance is not given due emphasis, even though the importance of the law of corresponding states as proved by van der Waals is sometimes mentioned for the programme leading to the liquefaction of helium.

The law of corresponding states is succinctly expressed by van der Waals.

> "If we express the pressure in terms of the critical pressure, the volume in terms of the critical volume, and the absolute temperature in terms of the absolute critical temperature, the isothermal for all bodies becomes the same. . . . This result, then, no longer contains any reference to the specific properties of various bodies, the "specific" has disappeared."[7]

Kamerlingh Onnes' theorem proves that the motions of the molecules of all substances, when in corresponding states, are dynamically similar, and

> All mechanical quantities, the derived absolute units of which can be given as powers of the fundamental units of length, mass and time, will be expressed in these systems of molecules by the same numbers, when measured in the system of absolute units, deduced from the fundamental units of length, mass and time belonging to each substance. And all such mechanical quantities of a substance in an arbitrary state, can be calculated from those observed with the other substance in the corresponding state by the ratio of the derived absolute units, in which these quantities are measured.[8]

Onnes' obsession with the study of molecular activity is not only guided by these considerations, that provide a heuristic rule which was to dominate the direction of this researches: his extensive use of analogy (not in any mechanistic manner), with the existing classical pictures, the use, that is, of this quite abstruse but liberating notion of analogy inherent in the law of corresponding states, becomes decisive in overcoming the limitations built in those situations which became the basis for his analogies.

Molecular theory, and more specifically van der Waals' law of corresponding states ("Elsewhere I have expressed this differently, in words that can be more easily understood: all substances form one single genus"[9]) introduce a new notion of elementarity and uniformity for all substances.

Kamerlingh Onnes' method expresses an approach which establishes, not so much the "hierarchy of substances" as is the case when we try to construct the world from each newly discovered or proposed level of elementarity and its relevant rules, but attempts to establish their uniformity, whereby information gathered for one substance under specific circumstances could be used to increase our knowledge of the state of another substance under more or less similar conditions. Kamerlingh Onnes had a very sound theoretical basis and motivation to embark upon a program to liquefy gases, and concentrated upon helium which, after the liquefaction of hydrogen, was the weakest link for establishing the "truth" of the law of corresponding states whose overall implications would be contributing towards the strengthening of the molecular hypothesis.

The fact that the investigations on molecular activity had such a primacy in Onnes' research programmes is not only evident from the frequent and explicit references to the consequences of many experimental results *vis-à-vis* the law of corresponding states, but also from the following instance, which is also quite characteristic. In 1912, one year after the discovery of superconductivity, there was a major "sum up" of the work done on the equations of state and there is a series of papers — particularly by Keesom — with further refinements of the equation of state whose corroboration would necessitate the performance of new experiments. Thus even though one would have "expected" that the importance of superconductivity would dominate over all the other activities, a new boost to the molecular physics branch was given, at a time when Leiden held the monopoly for superconductivity and when it seemed that further researches in molecular physics would just yield "corrections" to the proposed schemata.

The discovery of superconductivity, as momentous as it seems to us today — and obviously it was — did not have this impact on the community of physicists at that time.

> In trying to imagine the impact (superconductivity) made on physicists it must be remembered however, that the properties of the electric conductivity of metals was not well understood then, even at high temperatures. Also, in those days many other discoveries were made that could not be understood on the basis of existing theories. Superconductivity thus did not occupy a separate place.[10]

It is also important to acquire an insight about a situation not at all characteristic of Leiden's physics culture. The researchers working there were always extremely careful when making any generalizations and always tested them repeatedly before accepting their validity as a basis for further work. Nevertheless, they did accept the "conclusion" that superconductors were non-diamagnetic and did not bother to test it.

Finally, one has to realize that the liquefaction of helium was not simply a triumph in instrument-making and in perfecting techniques for going to ever-lower temperatures. It was also — and may have been more so — a triumph in pursuing in such a systematic manner the practical consequences of van der Waals' theory and in being guided for the setting up of such a program by the methodological implications of the theorem concerning the law of corresponding states proved by Kamerlingh Onnes himself.

ii. Work on magnetism in Leiden has not been adequately discussed and the need for such a discussion is threefold.

Firstly, to understand the deeper reasons for not discovering the diamagnetic character of superconductors before its discovery by Meissner in 1933 in Berlin, and together with the highly unusual situation for Leiden — where people were extremely careful when making generalizations — where a theoretical suggestion by Lorentz and a (wrong) interpretation of an experiment performed by Kamerlingh Onnes and Tuyn, discouraged people from using thermodynamic arguments.[11] And at the same time, we witness the daring proposals by Gorter and Rutgers to develop an approach for superconductivity depending on thermodynamic arguments together with a series of experiments which were progressively approaching the realization that superconductors are in fact diamagnetic.

Secondly, the "magnetic researches" which start very early and have their first success in the discovery of the Zeeman effect in Leiden, and have continuously added important contributions to low temperature physics, are closely interrelated with the "electrical researches" and contribute decisively to their success, and dominate the activities of the work in Leiden, especially after Kamerlingh Onnes' death in 1926.

Thirdly, due primarily to the "primitive" state of theory for the magnetic phenomena related to molecular physics, and the even "worse" situation concerning the theoretical schemata for low temperature magnetism, Kamerlingh Onnes together with his collaborators (who in the case of the magnetic researches involve people who have made important theoretical contributions) venture into extensive theorizing. They thus bring forth, in a systematic manner, methodological trends which, even though are present in his other researches, are not as cogently expressed as in the magnetic researches.

iii. A systematic study of the research programme(s) undertaken at Leiden during Kamerlingh Onnes' directorship, brings to the surface the methodological trends which were explicitly or implicitly expressed in the work of Onnes and which characterised Leiden's "physics culture". In attempting to summarize these trends, we shall term the overall approach — strongly influenced by positivist "prescriptions" — as sophisticated phenomenology: laws or theoretical proposals were very systematically tested, and the deviations from the predicted values carefully studied in order to be able to

> construct formulas describing the recorded results in a satisfactory manner. Explanations were then proposed and new experiments were planned to test the proposed explanations. When confronted with a new phenomenon there was a careful examination of the status of various theories, and the emphasis usually was not to provide novel theories or new explanations, but rather to bring forth all the facets of the new phenomenon. After all, Kamerlingh Onnes' motto "To knowledge through measurement" loomed high throughout his researches and especially in those instances when he was confronted with new phenomena.

We shall now consider in some detail the researches of Kamerlingh Onnes on the equation of state and on magnetism.

2.2. Researches on the equation of state

The law of corresponding states not only provided a motivation for Kamerlingh Onnes to pursue systematically his researches which led him to the liquefaction of helium, it is also the best means for comprehending certain characteristic methodological aspects of Kamerlingh Onnes' work: the law of corresponding states dictates a methodology whereby "analogy" becomes a dominating methodological rule in pursuing his investigations and establish very strict conditions for the extrapolation of existing theoretical schemata and the necessary modifications which have to be brought to the corresponding contexts so that they could accommodate these extrapolations.

In 1881 Kamerlingh Onnes published his 'General Theory of the Fluid State'[12] where, among other things, he proved a theorem about the law of corresponding states which had been proposed by van der Waals in 1880. It is interesting to note that not much significance seems to have been attached to this work, first published in Dutch, by other researchers working on questions concerning the law of corresponding states more than ten years after it was first published.[13] Van der Waals himself, however, duly acknowledges this contribution by Kamerlingh Onnes which was to be so decisive for Onnes' further researches.

> Kamerlingh Onnes, starting from his hypothesis that the corresponding states are states of similar mechanical movement, has *independently* demonstrated that the numbers expressing the proportionality of masses, of the lengths and of the time are determined by the molecular forces and the critical values $3\sqrt{u_k M}$ *and* $M_k^{5/3k}\, u^{1/3}\, T_k^{-1/2k}$.[14] (Emphasis added.)

The importance of this theorem for Kamerlingh Onnes' work is explicitly stated by him in 1894.

> I was induced to work with condensed gases by the study of van der Waals'

law of corresponding states. It seemed desirable to me to scrutinize the isothermal lines of the permanent gases especially of hydrogen at very low temperatures.[15]

In a very important paper in 1896, two years before the liquefaction of hydrogen by Dewar, Kamerlingh Onnes exploits to the utmost "*my theorem* concerning the law of corresponding states of van der Waals"[16] (emphasis added) in order to investigate the possibilities of cooling the hydrogen further by its own expansion. This he proposed to examine by using the law of corresponding states, and by actually working with more suitable substances at more suitable temperatures. He then used his theorem in order to predict from the data with existing apparatus, what was expected from the apparatus for the cooling of hydrogen, and applied the theorem to the experiments that had been already performed by Dewar.

In 1901[17] Kamerlingh Onnes developed in series the original equation of der Waals which was

$$p = (RT/(V - b)) - (a/V^2).$$

Onnes proposed

$$pV = A(1 + B/V + C/V^2 + D/V^4 + E/V^6 + F/V^8).$$

A is equal to RT for one mole, and A, B, C . . . are called the virial coefficients and depend on the temperature. For this dependence he proposed

$$B = b_1 + b_2/T + b_3/T^2 + b_4/T^4 + b_5/T^6,$$

where P, V, T = pressure, volume, temperature, respectively, and R, a, b = constants in the original van der Waals equation.

This equation, which assumed no quantum effects, was obtained by Kamerlingh Onnes by combining the observations concerning hydrogen, oxygen, nitrogen, and ether of Amagat with those concerning ether of Ramsay and Young, and those concerning isopentane of Young. The equation introduced the so-called virial coefficients which depended on temperature, and an expression for each virial coefficient was given. On the whole, the proposed equation has 25 parameters, with which the measured values could be described to a better approximation.

Kamerlingh Onnes had a well thought out "strategy" in constructing his equation of state:

I have followed in this communication a different method in considering the equation of state than has been done up to now. Various methods have been

> tried to empirically derive functions of u and t for Van der Waals' a and b by means of kinetic or thermodynamic considerations, but without obtaining a good agreement with the observations over the whole range of the equation of state. Neither was I successful in similar attempts which were repeatedly occasioned by my continued research on the corresponding states and other investigations resulting from them at the Leiden laboratory. Whenever I seemed to have found an empirical form, I discovered after having tested it more closely that it appeared useful only within a limited range to complete what had been found in a purely theoretical way by Van der Waals and Boltzman. Hence it appeared to me more and more desirable to combine systematically the entire experimental material on the isothermals of gases and liquids as independently as possible from theoretical considerations and to express them by series.[18]

All in all the equation of state in series was extensively used to test the law of corresponding states. Even though it did not lead to any noticeable improvement of the law of corresponding states, this procedure did involve many measurements to determine the second virial coefficient in order to achieve a deeper understanding of the deviations from the ideal gas law.

The investigations concerning the law of corresponding states were not confined to the improvement of the equation of state, and induced Kamerlingh Onnes to proceed to extensive measurements of the isotherms of both the diatomic and monatomic gases. The period 1900—1901 was the start of the systematic researches on the van der Waals ψ-surface on the isotherms of the diatomic gases and their binary mixtures. These, together with the measurements of the isotherms of the monatomic gases and their binary mixtures and the continuous improvements in the cryogenic apparatus eventually led to the liquefaction of helium. The law of corresponding states was again the guiding "spirit" for all these researches

> In sketching the meaning and extent of the work which is waiting for us. . . . I was guided by the law of corresponding states. Also our method of procedure is pointed out by this law.[19]

According to the van der Waals theory it is possible after a sufficient number of particular observations with mixtures of two known normal substances, to determine the constants which can allow us to construct the general equation of state for the mixture of these substances, and especially to predict the phenomena of condensation by the free energy (ψ) surfaces derived from that equation of state.

From the law of corresponding states one concludes that the following data are required for calculating the van der Waals ψ-surfaces for all temperatures:

a. One should have an equation of state for one normal substance with experimental measurements over the whole range of temperatures.

b. For the different mixtures of the two substances considered, as well as for these substances themselves, the deviations from the law of corresponding states must be known.
c. One must know the critical temperature and pressure of each mixture with the molecular proportion x of one of the components, derived from the law of corresponding states, as functions of those of the simple substances and of x.

These directions were fully utilised by Kamerlingh Onnes since

> ... with these data at our disposal van der Waals' theory will teach us all possible cases of coexisting phases of those substances if we roll tangent planes over the ψ-surfaces of each pair of substances for different temperatures.[20]

The researches went on to investigate how far, from an empirically correct representation of the isothermals (and thus of the ψ-lines) at different temperatures for a simple substance, one can find the ψ-lines for mixtures of different compositions at one temperature and, hence, also the unstable part of the ψ-surface.

A first step towards realizing this idea of Kamerlingh Onnes had been made by Keesom[21] who took for his basis the general equations by which van der Waals in his 'Theorie moleculaire' had expressed the relation of the critical quantities and the composition. He found the form of these equations for infinitely small x values and the equations of the ψ-surface in powers of x.

There is a series of crucial studies on thermodynamic diagrams which starts in 1898 and in effect continues through 1933. These are the first measurements on the system of isothermal lines near the plait-point, on improvements of the already developed open manometer and the measurement of the isothermal curves of hydrogen at 20°C up to 60 atmospheres. These follow the two series of the impressively systematic work on the isotherms of diatomic gases and their binary mixtures and of the monatomic gases and their binary mixtures.[22]

i. The work involving the isotherms of the diatomic gases and their binary mixtures can be divided into three categories, and work in all three was going on simultaneously.

a. Work which aimed explicitly at perfecting the instruments to be used for determining the isotherms of the diatomic gases and their binary mixtures[23] since improvements had been achieved both during the actual measurements or in the programs for the overall developments of the cryogenic apparatus. Here the main work was concentrated on developing the volumenometer in order to determine the compressibility of the hydrogen vapour and of the piezometers of variable volume for low temperatures.

b. Investigations on the isotherms and various other properties of hydrogen (density, critical point, compressibility of vapour).[24] There is the study of the isotherms of hydrogen from 100°C down to −238.3°C, and the determination of the critical point of hydrogen which was hindered by the difficulties involved at keeping these temperatures sufficiently constant for the time required to make the measurements — a difficulty which was overcome by the construction of the hydrogen vapour cryostat.[25]

The measurement of the compressibility of hydrogen vapour at and below the boiling point, and the successful calculation of the compressibility from an interpolation between the temperatures 2.2°K and 0.5°K led also to a calculation of the second virial coefficient, extending the region for which such a (reduced) coefficient had been known at that time (1913). In 1913 again[26] there was the determination of the liquid density of hydrogen between the boiling point and the triple point, as well as the very important discovery that solid hydrogen is heavier than liquid hydrogen, and therefore, on freezing, a contraction takes place in hydrogen of about 5% of the liquid volume.

Finally, a calculation by van Agt in 1925[27] who used all the then available data "proved" that the rule proposed by Kamerlingh Onnes and Keesom[28] for substances for which association is excluded, the critical temperature had an important influence on the whole network of isotherms *with regard to the law of corresponding states.*

c. The third category was the study of the isotherms, mainly of oxygen and nitrogen[29], of their vapour pressures[30] and of their behaviour according to the law of corresponding states. Both the isotherms as well as the vapour pressures were measured primarily for the purpose of testing the results against the "predictions" of the law of corresponding states, thus rendering it possible to compare the results with those of other gases. It should be stressed that in all these calculations and measurements extensive use is made of the equation of state in series developed by Onnes, neglecting the quantum effects, which are treated elsewhere.[31]

ii. For the isotherms of the monatomic gases and their binary mixtures the same pattern as for the diatomic gases is more or less followed, and one has helium instead of hydrogen, and argon and neon instead of oxygen and nitrogen.

The overall objective of the researches on the isotherms of the monatomic gases was to obtain a mean reduced equation of state by using exclusively the observations on the monatomic gases. There is, however, a difficulty related to substances with simple molecular structure, since the region that had been experimentally investigated extended over a small range of reduced pressure and reduced temperature. *If the law of corresponding states were strictly obeyed* this difficulty could have been overcome by reducing and then combining with each other the regions investigated for the various substances. Such a method was in fact used by Kamerlingh Onnes in 1908.[32] Concerning the objective set for the measurements of the monatomic gases, and unless, in the structure of the various atoms of the monatomic substances, further peculiarities were to be discovered which would influence the equation of state, the only influence exerted upon the form of the reduced surface was that of the critical temperature. This influence would have manifested itself in the deviations of the special equations from the mean equation, and in 1911 Kamerlingh Onnes together with Crommelin[33] using all the then available data, investigated the behaviour of argon with respect to the law of corresponding states.

1912 should be considered an important year for the researches on the equations of state in Leiden since, despite the fact that it was right after the discovery of superconductivity in 1911, one witnesses a boost for the thermodynamic researches. In a long article by Kamerlingh Onnes and Keesom in 1912 for the *Encyclopedie der Mathematischen Wissenschaften* titled 'Die Zustandsgleichung'[34] and summarising all the results from the researches on the equation of state and in a series of subsequent extremely interesting papers by Keesom[35], there were further investigations on the equation of state, by making use of Boltzmann's entropy principle, together with the studies on the second virial coefficient, and the then existing experimental data. The virial coefficients were derived both theoretically and experimentally, and the reliability of the final expressions depended on the assumptions made about the structure and motion of the molecules. Preferring the use of Boltzmann's entropy principle to that of Gibbs's method for canonical ensembles, with which it was shown to be equivalent, Keesom proceeded to obtain the equation of state of a single component substance.

Constructing specific processes for the determination of the macrocomplexion, microcomplexion and the equilibrium state, the second virial coefficient B was derived both generally, as well as for rigid ellipsoids of revolution which were subject to van der Waals' attractive forces; for material points (in the limit of rigid spheres of central symmetry) which exert central forces upon each other; for rigid smooth molecules of central symmetry having at their centres an electric doublet of constant moment; for rigid spherical molecules whose attraction is equivalent to that of a quadruplet placed at their centre; for rigid spherical molecules which besides collisional forces exert only Coulomb forces and for which the total charge of the active agent is zero, and for rigid spherical molecules carrying quadruplets. The papers of 1912 concluded by making an overall assessment of the experimental situation for both diatomic and monatomic gases *vis-à-vis* the theoretical improvements on the equation of state. The conclusions concern the types of dynamics implied by the various measurements considered for deducing the second virial coefficient, with the additional realization that for these measurements, the thermal behaviour of hydrogen approaches that of a monatomic substance, verifying, in effect, what Eucken had found for the caloric behaviour of hydrogen. Interestingly, viscocity coefficient measurements also supported such a conclusion. What, however, is of importance was that helium did not seem to fit into any of the above categories, since the maximum exhibited by the second virial coefficient for helium could not be derived by any of the above assumptions.

The measurement of rectilinear diameters has already been an indispensable aspect for the researches concerning the equation of state, and extensive measurements were done at Leiden for oxygen, argon, nitrogen, hydrogen, neon, helium and ethylene[36] and the results were found to be in fairly good agreement with theory.

The researches on the equation of state culminated in the liquefaction of helium on July 10, 1908 and we conclude this section by quoting from a speech by van der Waals when both he and Kamerlingh Onnes were, on November 19, 1908, awarded the highest mark of distinction of the Ancient Association of Physics, Medicine and Surgery for the "law of corresponding states" (van der Waals) and the "liquefaction of helium" (Kamerlingh Onnes).

> ... By carrying out measurements of pressure and volume at the lowest possible temperature, you were able to determine the so-called Boyle point i.e. the temperature at which, with very great volume, the substance follows Boyle's law. For all substances, the Boyle point is rather more than three times higher that the critical temperature, according to the equation of state 27/8 times temperature.
> ... Strictly speaking the question was now settled: helium too possesses this very same remarkable point.[37]

This is really a statement which, in retrospect, can be considered as closing one of the more dramatic periods in the history of physics: a period during which the validity of the molecular hypothesis was established beyond any doubt, and a period when the philosophical and theoretical implications of this hypothesis were systematically examined.

2.3. The magnetic researches

The study of the magnetization of liquid and solid oxygen and its compounds, the measurement of the susceptibility of nickel and various oxygen mixtures, together with the investigation of the properties of different sulphates and chlorides make-up the mainstream of the magnetic researches by Kamerlingh Onnes.[38]

What was primarily tested in these magnetic researches was the validity of Curie's law[39] for low temperatures, and this was followed by attempts to explain the observed deviations. Before we present the more important of these attempts we note two observations which were crucial in the further investigations of Curie's law.

i. Initial measurements classified ferromagnetic substances into three categories.[40]

The region of weak fields was identified by an almost total absence of hysteresis.

For regions of moderate fields the suceptibility changed very rapidly and hysteresis played an important role.

For regions of strong fields, magnetisation changed very slowly with changing fields. This work attempted to study whether all ferromagnetic magnitudes could be expressed as functions of the saturation magnetisation, especially since there was a belief that ferromagnetic initial susceptibility was due to a reversible turning of the direction of magnetisation — which was nearly the saturation magnetisation in elementary crystals. A knowledge, therefore, of the initial susceptibility could lead to a better grasp of the magnetic structure of the crystal itself.

The results obtained by Kamerlingh Onnes and Perrier led to the conception[41] that for all paramagnetic substances, or at least for one class of them, the deviations from Curie's law were governed by a law of corresponding states, the corresponding temperature for each substance to be taken as being proportional to a certain temperature characteristic of that substance. Later results[42] tended to confirm this. The general belief till 1914 was that, with the decrease of temperature, there were deviations from Curie's law, but always in the direction of decrease of susceptibility,[43] and Kamerlingh Onnes wondered whether the deviations may be attributed to the existence of a small zero point energy.[44] After the War, with the liquid helium experiments resumed, there were experiments to test Ehrenfest's hypothesis that at very low temperatures paramagnetic substances may show phenomena of hysteresis. This was not confirmed when measured for magnetic susceptibilities with high frequencies.[45]

ii. The susceptibility of oxygen suddenly became considerably smaller when oxygen became solid.[46]

For anhydrous ferrous sulphate, however, there was a maximum in the susceptibility as the temperature was lowered[47] and this supported the proposal by Kamerlingh Onnes that the deviations from Curie's law were intimately related to the energy properties of the Planck oscillators. There was, furthermore, the observation that the susceptibility depended on the magnetic field, and this led to the preliminary conclusion that, for these temperatures, substances which were ordinarily paramagnetic became ferromagnetic. Earlier, the case of saturation for ferromagnetic substances at high and very low temperatures had been studied by Kamerlingh Onnes and Weiss.

The understanding of the deviations from Curie's law for the low temperatures, especially after Langevin's derivation of the law, was the primary objective of many experiments in magnetism. The deviations first observed in oxygen for liquid hydrogen temperatures led to the proposal that they may be related to "how far the electrons which occasion magnetic phenomena are frozen fast to the atoms when the substance is cooled to very low temperature".[48] There was also the assumption that if similar deviations were observed for substances which obeyed Curie's law at ordinary temperatures, than "this freezing itself would lie at the very nature of paramagnetism"[49]. It should be noted that the idea of "freezing electrons" was extended to cover magnetic phenomena, after the proposals that such a mechanism may account for electrical resistance at low temperatures.

When, at about the same time, superconductivity was discovered and before abandoning this idea for good, Kamerlingh Onnes put forth the question of whether this freezing may have anything to do with the coming to rest of the Planck oscillators, and inversely, whether this coming to rest of the oscillators, accounts for the deviations from Curie's law. In 1907, however, there was Weiss's proposal whereby molecular magnetisation was expressed as whole multiples of the magneton[50], and the deviations from Curie's law were to be investigated with respect to this proposal and the possibility of discontinuous or continuous changes in the number of magnetons. The deviations from Curie's law in oxygen and the decrease of susceptibility as the temperature went down and oxygen changed states, gave rise to the idea that polymerisation, which may take the form of association in oxygen, may explain this behaviour of the decrease of susceptibility in terms of the distance between the molecules:

> We must not forget that it is by no means established that . . . the divergence from Curie's law . . . may be due to an association of molecules into complexes with a diminution of the number of magnetons . . . (and) whether the divergences . . . depend upon a peculiarity of the atom within a single molecule or from approach of the molecules up to a very small distance.[51]

Preliminary measurements of the susceptibility of oxygen versus its density gave some support to these considerations,[52] and when more careful experiments were performed a new set of hypotheses was put forth.[53] According to these one can assume the energy distribution of the molecules to be determined by quantum mechanics. Between these two extremes, Kamerlingh Onnes suggested that there may be "room for transitional" hypotheses, but he did not place any particular emphasis on the theoretical consequences of such an attitude. He continued, however, stating that "It is now of importance, not only for magnetism, but also for the law of molecular activity in general, to decide between these two different types of hypotheses by experiment".[54]

The most crucial point is to see what happens when paramagnetic molecules are brought to different distances from each other. If, in an experiment, it is shown that distances do not matter for susceptibility then "all hypotheses of the first sort (mutual influences) would, of course, fall to the ground. The measurements which should demonstrate this would be an *experimentum crucis*".[55]

The results of the experiments (measurements of the susceptibility of liquid mixtures of oxygen and nitrogen) showed that the deviations from the Curie–Langevin law found in the case of pure oxygen at low temperatures, were not an immediate consequence of the change of temperature, but were caused by the increase of the density, or equivalently, by the distance between the molecules becoming smaller. Using the relation $\chi\ (T + \Delta) =$ constant (Δ being a parameter depending on concentration) Kamerlingh Onnes and Perrier reached the conclusion that the change in density of oxygen alters only the specific magnetisation without changing the Curie constant.

Any attempted theoretical interpretation, then, should have accounted for the following two facts: the change in magnetisation with density and the parallelism of the lines $1/\chi = f(T)$. Langevin's theory, supplemented by the hypothesis of the negative molecular fields turned out to be sufficient to explain these facts, i.e.,

$$1/\chi = T/c + N\alpha \qquad (N = \text{coefficient of molecular field})$$

gives $(T + \Delta) = c$ where $\Delta = cN$, and which agrees with the experimental results if one assumes that Δ (or $N\alpha$) decreases with decreasing density. The first question is whether the field is equal to a particular power of the distance between the molecules. If we assume that $N = \rho\alpha^n$, then the molecular field is of the form $\rho\alpha^{n+1}$, and at constant temperature $1/\chi = f(\alpha)$ is a parabola. The experimental measurements seemed to verify this, and thus supported the hypothesis that the molecular field of oxygen changed proportionally with density. If, however, oxygen has a negative field, then the proposed "law" would differ totally from Weiss's law that he derived with positive field for alloys of ferromagnetic metals, and from which he inferred that the influences between molecules are expressed by the inverse sixth power of their distance, whereas the formula used by Onnes and Perrier would lead to an inverse third power.

> At present, we need not see any contradiction between these two results, as the conditions for which the two laws of distance hold good, are quite different. This applies both to the nature of the substances and to the state of aggregation in which they were examined. Moreover, it must be particularly borne in mind that the sign of the molecular field is different in both cases. The part of the curves referring to the change of N with the concentration, which Weiss makes use of in his theory, lies entirely in the positive fields, the transition to negative fields is curved. We are in complete ignorance as to the origin of the mysterious influences which cause the phenomena ascribed in the molecular field. There is no ground, therefore, to expect that both fields are subject to the same law. Should it be confirmed that both kinds of molecular

> field depend upon the distance of the molecules according to different laws, we might even see in this a proof that in both cases influences are at work which are the effect of different causes. . . . Although the hypothesis of the negative molecular field is sufficient to describe the phenomena, it is not devoid of interest to consider in how far the other hypotheses can be reconciled to the observations.[56]

Yet again we witness Onnes' characteristic methodology at work:

Testing a law; preliminary assumption(s) to account for the deviations; planning of the relevant experiments to decide on the plausibility of the assumption(s); precise measurements; (new) assumption(s) and the proposal of phenomenological formula(s) which describe the data satisfactorily; emphasis on the dangers which are entailed if one accepts these formulas as the theoretical explanation of the phenomenon; the status of other hypotheses *vis-à-vis* experimental results; planning of new experiments.

2.4. Concluding remarks

What has been attempted in this chapter is the description of various methodological aspects of the work of Kemerlingh Onnes and his closest collaborators. Kamerlingh Onnes is primarily known for liquefying helium and for his discovery of superconductivity. What is shown in the chapter is that the monumental experimental contributions of Onnes involved a considerable amount of theorising and they were the result of a carefully planned long-term research strategy.

Faithful to his motto "knowledge through measurement" Kamerlingh Onnes developed a sophisticated phenomenological attitude while remaining within the strict bounds of the positivist tradition. This attitude was developed not only to discourage premature generalisations or hastily announced new discoveries, but, mainly, in order to be able to investigate, in a thorough manner, as many features as possible of the phenomenon being studied. It is in this context that one will have to assess Kamerlingh Onnes' attempts to provide theoretical explanations to various phenomena: he is never really interested in the explanatory schemata as such, but only in so far as these schemata can facilitate the further investigations of the different facets of what is being studied. There are many instances when various hypotheses are proposed for the tentative explanation of a particular phenomenon (or behaviour) and even though the experimental data can be considered as not excluding (and at times even strengthening) a particular hypothesis, the concluding remarks in these papers do not overemphasize this point. Almost always new hypotheses are being proposed and new experiments suggested to test them in this continuous process of bringing to the surface ever newer features of an observed phenomenon.

Kamerlingh Onnes' reformulation of the law of corresponding states guided his researches on the equation of state which eventually led to the liquefaction of helium. This achievement — even eight years after Planck's proposal of the quantum hypothesis — was one of the more spectacular manifestations of the molecular hypothesis and it displayed the "strength" of the methodology implied by the mechanistic world outlook taken to its limits. A few years later, Onnes discovered that most "extreme" quantum phenomenon, superconductivity, and subsequent attempts at Leiden to explain this phenomenon established a characteristic methodology which dominated all future developments of low temperature physics: to form new (non-classical) concepts out of the classical framework while, at the same time, trying to form the new conceptual framework implied by the relative autonomy of these new concepts. It was none other than the process of concepts out of contexts. In the work of Kamerlingh Onnes one witnesses the foundations of this complex process which was in a way completed in the mid 50s by Bardeen, Cooper and Schrieffer and Feynman and Onsager. The next chapters present these developments analytically.

CHAPTER 3

Superconductivity: the paradox that was not

3.1. The background

The earliest attempts to formulate a theory of electrical conductivity of metals brought into prominence the variation of conductivity with temperature. As lower and lower temperatures were reached toward the end of the nineteenth century, physicists began to study the resistivity of metals as a function of temperature.

Simultaneously with the efforts for the liquefaction of hydrogen, experiments were being performed at the temperature range of liquid air and, in fact, Dewar and Fleming (1893) carried out an extensive series of such measurements. Their early results tended to confirm that "the electrical specific resistance of all pure metals will probably vanish at the absolute zero of temperature".[1] After the liquefaction of hydrogen it became possible to continue these measurements for considerably lower temperatures, and it was found that "Instead of continuing their straight downward course, the resistance curves bend round, indicating the survival, at 0° absolute, of a finite value for this property. An emphatic warning was thus conveyed against trusting the continuity of change".[2]

Nevertheless, even by 1901, as Dewar (1901) noted, the form of the law correlating electric resistance to temperatures of the liquid hydrogen range was still unknown despite the successful theory of the electrical (and thermal) properties of metals proposed by Riecke (1898) and Drude (1900). They treated the electric current in a metal as a drift of an electron gas under the influence of an electric field. The conduction electrons in such a model move freely in the spaces between the heavy, fixed atoms of the metal, with which they exchange energy by collisions and so they contribute towards the establishment of thermal equilibrium. The electrons were considered to be free, and, apart from the collisions with the atoms, they were assumed to behave as an ideal gas, their mutual interaction being neglected. If an electric field is then imposed, the motions of the electrons will no longer be entirely random and an electric current will be set up in the direction of the electric field. Ohm's law, then, for the conductivity σ is found to be:

$$\sigma = \frac{e^2 n\lambda}{2mv},$$

where e and m represent the electronic charge and mass, n the number of free electrons in unit volume, λ the mean free path and v the root-mean-square electron velocity. If one now replaces the mean kinetic energy $1/2\ mv^2$ by $3KT/2$ — the value ascribed to it by classical kinetic theory — Drude's expression for σ becomes:

$$\sigma = \frac{1}{6}\frac{e^2 n\lambda v}{KT}.$$

A temperature gradient in a metal will also cause an electron current which is calculable by the same principles. Drude's expression for the coefficient of thermal conduction is $u = hv\lambda k/2$, and the ratio of u to σ becomes:

$$\frac{u}{\sigma} = 3\frac{K^2}{e^2}T.$$

This expression was in agreement with the old empirical law of Wiedemann and Franz (1853): the ratio of thermal to electrical conductivity is the same for all metals at the same temperature. The derivation of this law represented one of the striking successes of the early form of the electron theory of metals. This successful prediction was to a certain extent offset by the difficulty in reconciling the expressions of σ and u separately with the experimental data. Lorentz (1905) started his investigations by using the statistical theory devised by Maxwell and Boltzmann and also investigated the dynamics of the collision process more carefully. His results differed a little from Riecke's and Drude's and the attempts to adjust the parameters to fit the data produced serious anomalies in all the other expressions. In order to obtain the correct temperature variation of σ and to reproduce the correct numerical value of the ratio of u and σ the product $n\lambda v$ must be independent of temperature. Since, however, by assumption $\nu \sim T$, the free electron theory required $n\lambda \sim 1/T$.[3] But if this was the case there could be no accounting for the rapid fall of resistance at extremely low temperatures; a decrease proportional to T itself was quite insufficient. The low temperature regions became a fundamental difficulty for the electron theory of electrical conduction.

Although neither the Riecke-Drude theory, nor its modification by Lorentz seemed able to account for the experimental results at very low temperatures, the investigations by Kamerlingh Onnes and his collaborators between 1904 and 1908 seem to have been at least partly motivated by Lord Kelvin's (1902) proposal. This proposal followed from regarding the electrons as a substance characterized by an equation of state: as temperature goes down, the resistance will increase greatly, after passing through a minimum, due to "electron condensation" — the freezing of the electrons on the atoms.

Dewar's resistance measurements in liquid hydrogen temperatures had indicated that the resistance of some metals dropped at a conspicuously slow rate and Kamerlingh Onnes (1904) did acknowledge that Dewar's researches gave credence to Lord Kelvin's proposal.

By 1906 Kamerlingh Onnes had an efficient hydrogen liquifier and, with J. Clay, investigated problems of thermometry at liquid hydrogen temperatures in order to reproduce and extend Dewar's measurements. They embarked upon a program in order to establish

> the existence of the point of inflection which may be *expected* in the curve representing the resistance as a function of the temperature, especially with regard to the supposition that the resistance reaches a minimum at very low temperatures, increases again at still lower temperatures and even becomes infinite at the absolute temperature 0°. . . . And this has been done especially because temperature measurements with the resistance thermometer are so accurate and so simple.[4] (Empasis added.)

Onnes and Clay (1906a, b, 1907a, 1907b) estimated the values of the "points of inflection" of platinum and gold, and concluded that for metals "in the purest and normal state the point of proportionality lies probably still below the temperatures which are to be reached with liquid hydrogen".[5]

Their observations showed that if one wished to take account of the resistance over the whole region of low temperatures, one would have to devise rather intricate formulae. Kamerlingh Onnes and Clay (1908), and Clay (1908) tried to improve their formulae and to extend their investigations by studying the influence of small admixtures on the change with temperature of the electrical resistance of pure metals. The extrapolation of their data to lower temperatures suggested the onset of a constant residual resistance in the helium temperature range, whose value was lower the purer the specimen became. Moreover, the fact that the expected minimum value for the resistance "as regards temperature, always lies beneath the limit of observation as yet . . . excite a doubt of Lord Kelvin's opinion, that the resistance at the absolute zero point does not become zero, but may have a very large value . . .".[6] But Onnes soon abandoned his tentative belief in Lord Kelvin's proposal after the first investigations concerning the resistance of metals at helium temperatures.

3.2. The Discovery of superconductivity and the first attempts to explain it

On the 10th July, 1908, Kamerlingh Onnes (1908) liquefied helium and tried unsuccessfully to solidify it by evaporation under its own vapour pressure. In 1909 and 1910, Kamerlingh Onnes repeatedly attempted to solidify helium, failed again, and decided to concentrate on the measurement of various physical parameters in the newly accessible temperature range. One of his main projects was to determine the electrical resistance of metals as a function of temperature. Even the very first results obtained at helium temperatures displayed the inadequacy of Lord Kelvin's theory of conduction: the resistance of very pure platinum became constant[7] instead of passing through a minimum or of tending to vanish at absolute zero "as would be expected, if the electrons which provide the conduction were firmly frozen to the atoms at low temperatures".[8] Onnes attempted to provide a theoretical explanation by ascribing the constant value of the resistance to the presence of impurities. He was, then, led to expect that pure metals could show a practically infinite conductivity. According to this result the dominant views on electric conduction had to undergo a fundamental change: it was not the free electrons that freeze in the metal; on the contrary, it was the impediments which oppose the rectilinear motion of the free electrons, which then lose their "extension".

Since it seemed that it was only impurities that prevented the resistance of platinum and gold from disappearing, Kamerlingh Onnes decided to carry out an experiment with "the only metal which one could hope to get into wires of a higher state of purity viz. mercury".[9] He was fortunate in his choice because mercury turned out to be a conductor with the property which became known as superconductivity — a property not shared by all conductors whatever their purity. At the same time as he was experimenting with mercury, Onnes determined the variation with temperature of the resistance of pure samples of gold, and found that it fell gradually to a value smaller than he could measure. More exact measurements later showed that the resistance did not vanish completely, and gold was not a superconductor.

The preliminary mercury results seemed to confirm Kamerlingh Onnes' (1911c, d) expectations. However, the fit of the curve representing resistance as a function of temperature between the melting point of hydrogen and the boiling point of helium was not as good for mercury as it was with gold. The change in resistance took place faster than the rate of change predicted by a formula proposed after the measurements on platinum,[10] and the experiments were repeated. The results left no doubt about the disappearance of the resistance of mercury. The temperature at which the resistance seemed to vanish "was found to be slightly more that 4.2 °K".[11] But the disappearance did not take place gradually but *abruptly*, and this sudden fall could not be foreseen by Onnes' "expectations" concerning the behaviour of resistance at very low temperatures. In November 1911 the phenomenon was reaffirmed at 4.19 °K. These measurements showed that from the melting point of hydrogen to the neighbourhood of the boiling point of helium, the curve exhibited the ordinary gradual lessening of the rate of decrease of resistance. This was practically the same as the result given by the formula derived by fitting the previous set of data. "A little above and a little below the boiling point of helium from 4.29 °K to 4.21 °K, the same gradual change was clearly evident . . . but between 4.21 °K and 4.19 °K the resistance diminished very rapidly and disappeared at 4.19 °K".[12]

Experiments to measure the resistance of mercury at helium temperatures were repeated in 1913, and the emphasis was now somewhat shifted. What was studied was "the potential difference necessary for the electric current through mercury below 4.19 °K".[13] The phenomenon of the sudden drop of resistance was firmly established, it was realized that impurities did not play any role — at least in the case of mercury — in hindering the disappearance of the ordinary resistance, and the phenomenon for the first time was called the "superconductivity of mercury".[14]

At the time Kamerlingh Onnes decided to investigate the variation of resistance in the newly available range of low temperatures below 4.2 °K, the theory of electrical conductivity was in a rather rudimentary state, and any of the three modes of behaviour sketched in Figure 2 seemed possible.

Thus, *a* would occur if the resistance was due entirely to the obstruction of electronic paths by thermal vibrations, *b* would occur if obstruction by impurities and imperfections was important, while *c* would occur if the number of free electrons available to carry the current fell off rapidly at low temperatures due to their "condensation" on the atoms.

The first measurements of electrical resistance at liquid helium temperatures in 1911 "brought quite a revelation". The implication, that pure metals would show infinite conductivity, not only undermined the notion of electrons' freezing to atoms, but also, according to Onnes, suggested that thermal agitation of Planck's oscillators might provide the mechanism "responsible" for electrical

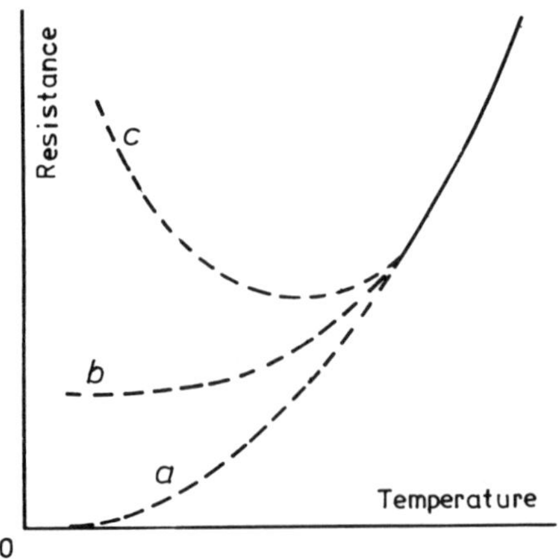

Fig. 2. Possible forms of temperature variation of resistance (schematic).

resistance: "At the lowest temperature therefore the conduction electrons are not bound, *but the factors which hinder their movement disappear.* The obvious thing was to look for these hindering factors in the energy of Planck's vibrators".[15] (Emphasis added.)

Planck (1911a), in his quantum theory of radiation, had introduced oscillators whose mean energy vanishes at a temperature which is very low, but distinctly above absolute zero. Einstein (1907) had established the new theory of specific heat, which regarded the heat-agitation of solids as the vibration of these oscillators. Kamerlingh Onnes realising the usefulness of these ideas for explaining the behaviour of resistance at low temperatures was claiming that

> If we accept that the impediments to the movement of the free electrons, through the metal, depend on the amplitude of these vibrators, then we get a simple view of the variation of the electric resistance with the temperature and the practical vanishing of it when the vibrators are frozen, is explained.[16]

The temperature at which the mean energy of an oscillator vanishes depends, according to Planck, on its frequency. By deducing the frequency of the mercury oscillators from the law of corresponding states as applied to metals, "it could be foretold, that the resistance of a wire of solid mercury . . . would fall to inappreciable values at the lowest temperatures which I would reach. With this beautiful prospect before me there was no more question of reckoning with difficulties".[17] At 1/10 of the temperature determined by the frequency of the mercury vibrators, Kamerlingh Onnes took the energy to be practically zero.

Actually, Kamerlingh Onnes' formula for the resistance of metals, proposed in his paper on the resistance of platinum at helium temperatures, was the result of a curious mix of classical and quantum ideas. Calculations on this basis for platinum, silver, gold, and lead did produce a qualitative correspondence with the measurements of the purest samples available down to 4.3 °K.

At about the same time as Kamerlingh Onnes' experiments concerning the electrical resistance of conductors at helium temperatures, the measurements by Nernst (1911), Eucken and Lindemann of specific heats as a function of temperature at liquid hydrogen temperatures came out higher than the values predicted by Enstein's theory of specific heat. While trying to account for this discrepancy, Nernst was struck by the remarkable analogy with the case of temperature dependence between specific heat and specific resistance and proposed an empirical formula for the resistance suggested directly by the form of the Planck radiation formula.[18] Lindemann (1911) also proposed a formula for the variation of resistance at high temperatures (higher than 14 °K). This formula, however, gave worse agreement with the measured values than Onnes's at temperatures lower than 14 °K.

To test his formula further, Onnes chose mercury for some of his subsequent experiments because he could get it most easily in a state of high purity. The result of the experiments left no doubt about the "disappearance" of the resistance of mercury at 4.19 °K. "At this point within some hundredths of a degree came a sudden fall, *not foreseen by the vibrator theory of resistance that I had framed* . . .".[19] (Empasis added.)

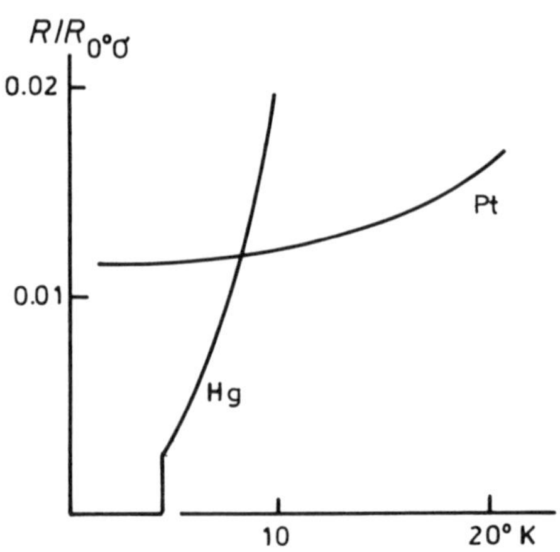

Fig. 3. Variation of $R/R_{0°C}$ (ratio of the resistance to resistance at 0°C.) of platinum and mercury.

In an account of his work on mercury to the first Solvay Congress in 1911 Onnes (1912) said that his resistance formula — resulting from a combination of Planck's considerations and the Riecke—Drude—Lorentz theory — was nothing more than a rough tentative application of the theory of quanta, and that it could not explain the way the resistance vanished. In the discussion following Onnes' presentation, Langevin asked whether "the very fast variation of the conductivity of mercury near 4 °K corresponds to a change of state and whether it may be accompanied by any observable structural modification of mercury".[20] Experiments with the object of settling this point were immediately planned.[21] It was proposed that "should there exist such a new modification, it would differ from ordinary mercury at higher temperatures chiefly by the property that the frequency of the vibrators in the new state has become greater, and therefore the conductivity rises to the extremely large value exhibited below 4.19 °K".[22]

In his report to the same Congress, Nernst (1912) discussed a revised specific-heat formula, published by Lindemann and himself (1911) which offered a possible explanation for the rapid change in frequency that Kamerlingh Onnes needed for the explanation of the experimental measurements. Planck (1912) discussed his concept of zero point energy and gave a new formula for the mean energy of an oscillator. The zero point contribution implied an energy of vibration much too high to be reconciled with the data and Onnes pointed out this difficulty in the explanation of the resistance data in terms of oscillators.

Onnes came to regard the apparent breakdown of Ohm's law below the transition temperature as a consequence of the increase of the distance travelled by electrons between two collisions to a size comparable to the dimensions of the conductor. The drift velocity under the applied field could not then be neglected when compared to the velocity of the "heat movement" and thus violate a fundamental assumption of the simple electron model. He proposed an essentially different hypothesis concerning the movement of the free electrons in the superconducting state: the movement of electrons is carried on by the current for considerable distances, but each separate electron only moves one modecular distance. An electron "jumping across on to an atom of the superconductor from one side causes an electron on the other side of the atom to jump into the next atom, etc. ... The migration speed is thereby propagated through the superconductor without the performance of work".[23]

According to this hypothesis the Ohm resistance above the vanishing point was due to the action of the vibrators which bring the atoms to such a distance from each other "that the electrons cannot jump from one atom to another without doing work ...".[24] Whether one accepted that the free path was continuously described by the same electron or that it was broken by the movement being transferred over a distance from one electron to another, a difficulty arose because of Planck's zero-point energy. Planck's new formula for the mean energy of an oscillator implied that the distance travelled by an electron was determined only by that part of the energy of the vibrators, which was dependent on the temperature. When the excess energy above the zero point energy fell to the small value which corresponded to the vanishing point, the resistance suddenly became zero.[25]

Onnes went on to offer an explanation for yet another "peculiar" property of the superconductors: resistance was found to be restored when a relatively high electric current passed through the superconductor. The generation of heat was the obvious proposal for a preliminary explanation of this phenomenon, but since no such heat was detected, he assumed that the "less conductive" particles were separated out of the mercury[26] during the freezing or they came among the mercury crystals in some other ways, thus introducing a resistance into the path of the current. This was a disappointment since part of Kamerlingh Onnes' enthusiasm about superconductivity was about the prospects of using superconducting wires with very intense currents through them in order to be able to do experiments with very high magnetic fields.[27] Establishing at the same time that mercury, even in an inpure state, could still be superconducting, Kamerlingh Onnes (1913d) concluded that for a thread of mercury there was a residual resistance called "microresidual" resistance distinct from the "additive mixture" resistance attributed to impurities, and that this resistance was responsible for creating various Peltier-sites in the threads of mercury which were "not too unevenly distributed".

Kamerlingh Onnes had time for only one more set of experiments on superconductivity before the First World War essentially forced a halt to all work with liquid helium. The objective was to pin down a still more accurate value for the micro-residual resistance below the transition point by measuring the time constant in a superconducting circuit. The first attempt was made in April of 1914[28] and he stated, interestingly, that "a deep impression is made by the very striking realization which it gives of the mechanism imagined by Maxwell completed by the conception of the electron".[29]

Three of the attempts, during the following years, to find a solution to the problem of superconductivity are of interest to us. In 1915 J. J. Thomson claimed that superconductivity "is another, and fatal objection to the theory that metallic conduction is due to the presence in the metal of free electrons which drift under the electric force"[30] and that the effects discovered by Kamerlingh Onnes were in accordance with his old "corpuscular theory of matter". In this line of thought he made an attempt to explain superconductivity by a method similar to that by which "Weiss explained the existence of permanent magnetism below a critical temperature".[31]

Thomson's approach appealed mainly to people who were critical of quantum theory, but did not appeal to Kamerlingh Onnes. Lindemann (1915)[32] rejected the concept of a gas of free electrons and supposed that the electrons in a metal are so coupled as to constitute a lattice of considerable rigidity. Hence conductivity was represented by the motion of an "electron lattice" through the ionic lattice of the metal and the superconducting state of a metal occurred at temperatures where the thermal agitation of the ionic lattice had decreased so much as to offer no appreciable interference to the free motion of the electron lattice.

In another theory, proposed by Benedicks (1916), and also considered by Bridgman (1917a, 1917b), conduction was ascribed neither to an electron gas nor to a lattice of coupled electrons, but to the passage of valence electrons from one atom to the next. For certain metals a point is reached, by lowering the temperature, where the outer electron or bits of neighbouring atoms come into tangential contact. Below this point the radiationless, and therefore, resistanceless bits form a continuous chain along which the electrons would move under the influence of an external field.

In 1921, Kamerlingh Onnes once again, for the first time since 1914, tackled the problem of superconductivity. In the third Solvay Congress he gave two papers: The first was about paramagnetism in low temperatures and the second was titled "Les superconducteurs et le modele de l'atome Rutheford-Bohr".[33] In this paper he puts forth a set of questions which, according to Onnes, deserved a systematic investigation during the coming years.

1. Since the Rutherford—Bohr atoms unite to make a metal, what happens to their electrons? Do they lose all or only part of their kinetic energy?

2. How many kinds of electrons can we distinguish in a metal ("free" or "fixed" in one way or another)? What is the statistics of their movements according to quantum theory?
3. Are the movements of the conduction electrons of adjacent atoms coherent?
4. What is the mechanism with which the conduction electrons in the ordinary conductors transmit to the degrees of freedom of the thermal agitation the energy they have accumulated from the external electromotive force?
5. Is there, in all these, a physical velocity of the conduction electrons which may play any role?
6. How is it possible for atoms to have superconducting contact, how can they form superconducting filaments which can open up a macroscopic route so that the conduction electrons can pass, without transmitting their energy to the thermal degrees of freedom of the surrounding atoms, and how is it that the conduction electrons are guided to follow this route among the atoms which are in thermal movement?
7. Within what limits of the changing circumstances does the superconducting route resist a "rupture"?

Up to now we have talked about the cases where there is no external magnetic field. The magnetic influence intervenes naturally in the solution of all the questions we formulated. Introducing the external field every question is doubled. And others would follow. We would be engaged in a very vast field where almost all our experiences for the superconducting state are faulty. There is nothing except the Hall effect for which there have been researches and which have shown that the electromotive force — observed in an ordinary manner — disappears with the resistance, so long as there is disappearance of superconductivity as well.
We confine ourselves to adding the following question,

8. What is the reason for the equivalence of the magnetic field and of the temperature for the destruction of the stable adiabatic flux of the electrons during the agglomeration of atoms a large number of which is in thermal movement?.[34]

There is a deep insight in these questions, and the importance of the magnetic considerations is so much empasized for the first time.

The researches on the Hall effect and the study of the resistance with the change of an external magnetic field had been started in Leiden by Labret and van Ederdingen as early as 1895. Right after the discovery of superconductivity, Kamerlingh Onnes himself became involved in these researches, which apart from determining the Hall coefficient for various substances under various conditions, acquired an added significance for the following two reasons.

i. The investigations on the Hall effect were studies of the conduction mechanisms in metals — something that Kamerlingh Onnes (1912) emphasized right in the first paper of the series.
ii. These investigations gradually led to one of the most important discoveries concerning the superconductivity of metals: the fact that superconductivity is destroyed by the application of a large enough external magnetic field. After a suggestion by J. Becquerel (1912) that the Hall coefficient could vary linearly with the applied magnetic field and that for strong magnetic fields the resistance varies directly with the external field, experiments did in fact show (for e.g. bismuth) this to be the case at liquid hydrogen temperatures. The discovery of this property was in the last paper of the series (Kamerlingh Onnes 1914a) and in the later papers one had the explicit study of this newly discovered property.

Kamerlingh Onnes measured the Hall coefficient for bismuth, gold, silver, electrolytic copper and palladium.[35] Bequerel proposed a formula for the Hall coefficient, $R_H = a' H + b'$, where H is the applied field, and at strong fields this formula predicted the resistance to be linearly dependent on the field. There was further testing of this linear dependence of the Hall coefficient with the magnetic field. In 1912, it was established that for bismuth the resistance at liquid hydrogen temperatures was increased with the increase of the magnetic field. For high fields, the resistance was increased as the temperature was lowered.[36] Finally, what was inferred from the extrapolation of the data from the Hall effect in hydrogen temperatures, was that not even large magnetic fields would induce a small resistance. In 1914 the opposite was discovered (Kamerlingh Onnes 1914a), whereby an external magnetic field disturbed superconductivity by "generating resistance" in lead and tin. Onnes characteristically emphasized not the failure but the novel property: "An unforeseen difficulty is now found in our way, but this is well counterbalanced by the discovery of the curious property which is the cause of it".[37] The researches on the Hall effect were completed with the discovery of this property, and which followed the discovery that superconductivity is destroyed when current above a certain threshold value passes through the superconductor. Kamerlingh Onnes was inclined to believe that these two phenomena were related, and he proposed a tentative explanation.

> If it were to be proved that the vibrators which cause the resistance can only be set in motion when the stream of electrons passes them with sufficient rapidity, then it would not be surprising that the magnetic resistance does not arise until the rapidity of the circulating motions of the electrons is great enough to carry the atoms with it and set them in rotation, by which they can then disturb the regular motion of the electrons.[38]

Despite the use of the quantum mechanical notion of zero-point energy, the overall approach was essentially mechanistic. Researches concerning this phenomenon continued after the war when it became possible to have liquid helium again, and the effort was mainly concentrated — but not exclusively — on testing Silsbee's (1916) hypothesis. This hypothesis stated that the threshold value of the current was equal to that value at which the magnetic field, caused by the current at the surfaces, was equal to the magnetic threshold value. Silsbee's hypothesis was important in that, if it turned out to be correct, then one would have had to investigate only the magnetic disturbance.[39] This partly explains why there were no extensive measurements at Leiden on the disturbance of superconductivity by increasing electric current. Data, primarily from tin, were not in disagreement with the hypothesis of Silsbee, and further experiments indicated that the threshold current was of importance only because of the critical field it causes.[40] It was, furthermore, observed that the restoration of resistance did not take place suddenly and this — in a note added to the Communications, and which was not in the original Dutch text — was attributed, probably, to the hysteresis effect.

By 1925 there was no satisfactory theoretical explanation for the superconducting state of metals, and new thoughts were expressed as to how the researches should continue, "which might serve as a guide for further investigations, and it seems desirable to try by changing the external conditions, to discover the factors which play a role in the appearing of the phenomenon".[41] Such investigations started during the last year of Onnes' life, and they showed that elastic deformation caused a small displacement of the temperature at which a metal became superconducting. The straightforward continuation of these studies was to see whether the threshold value of the magnetic field might be influenced by such a mechanism. When deformation vs. magnetic field was measured and the expected displacement found, "a quite new phenomenon was discovered which gave an unexpected turn to the inquiry".[42] The magnetic transition curve (transition from the superconducting to non-superconducting state) was a hysteresis curve. That is, if the strength of the magnetic field is increased, the resistance comes back at a higher strength of field than that at which it disappeared in a diminishing magnetic field. Concerning these researches, a hypothesis was put forward that if extension and compression of the superconducting metals may be considered as equivalent to an increase and decrease of the distances between the atoms, then the results found for tin supported the idea that a relatively large space between the atoms became favourable for the appearance of superconductivity — an assumption which was not verified by subsequent measurements.[43] It should finally be mentioned that as early as 1922 Kamerlingh Onnes and Tuyn had remarked that "regarding a difference of vanishing point temperature (for resistance) for isotopes it seemed not impossible that the occurrence of superconductivity might be influenced by the mass of the nucleus".[44] This guarded statement was not exploited at all by subsequent researchers who, until 1950, believed that there was no dependence of the critical temperature on the isotopic mass, at which time experiments showed that the critical temperature varied as the inverse square root of the isotopic mass.[45]

Was superconductivity a change of state of the electrons? Indeed, physicists had been looking for a possible phase transition, but preliminary measurements that had been done to ascertain whether there was a heat of transition at the transition point led to negative results, and X-ray interference experiments did not reveal any change in the crystal structure (Keesom and Kamerlingh Onnes 1924). There was no change of density and no latent heat at the transition point. So they assumed that there was no phase transition and looked for a mechanism to explain an infinite free path in an essentially unmodified electron distribution.

It was a misconception about the magnetic properties of superconductors which became a major stumbling block for any progress. In 1919 Lippman had clearly formulated the idea that the magnetic flux linked by any closed loop inside a superconducting material cannot change and in 1924 Lorentz drew attention to a remark originally made by Maxwell concerning perfect electrical conductors. If a conductor has no resistance there will be no electric field inside it even when a current is flowing. Maxwell's second equation of the electromagnetic field then becomes $d\mathbf{H}/dt = -c \text{ curl } \mathbf{E} = 0$ and on integrating it, we find,

$$\mathbf{H} = \mathbf{H}_0, \tag{1}$$

where $\mathbf{H}_0$ is the field which was in the specimen when it lost its resistance. Thus, as long as the specimen has perfect conductivity, the field distribution inside cannot be changed by any external changes and can be regarded as "frozen-in". The physical meaning of this result was simply that any change of external magnetic field induced currents on the surface of the metal, and the magnetic field of these currents inside the metal compensated the change of external field, thus keeping the field inside the metal constant. Since there is no resistance, such surface currents cannot die away, and so the field inside the metal remains constant with time. This physical assumption was regarded as being so self-evident as not to require an experimental test. Thus, in the early literature on superconductivity frequent references to the supposed "frozen-in" fields can be found, although their existence had never been shown experimentally. Superconductors were not regarded as being diamagnetic and thus it became impossible to use thermodynamic arguments since the latter required reversibility which, however, was not allowed by the assumed diamagnetic character of superconductivity.

The rule of Lippman explained many features of the behaviour of superconducting circuits. But it was thought that, if one were to cool a solid body, for instance a sphere, in a magnetic field, the same would happen; the field would be unchanged at the transition and it would retain the value of the field in which the superconductor was cooled down, even if the external field is switched off. Kamerlingh Onnes (1914) and Tuyn's (1929) experiment[46] with a lead sphere seemed to confirm this idea, but the sphere was hollow, and if one used hollow spheres or a sphere with some imperfections there would indeed be a frozen-in field, although smaller than that predicted by the simple theory of frozen-in fields.

The difficulties of the electron gas theory of metallic conduction were precisely those which occurred in several other classical problems where the use of the Boltzmann statistics led to disagreement with observation. The development of modern quantum theory led, in 1926, to the introduction of the Fermi—Dirac statistics, based upon the exclusion principle (enunciated by Pauli in 1925), and which was applied by Sommerfeld (1928) with considerable success to the electron gas in a metal. Sommerfeld showed that many of the most serious difficulties of the Drude—Lorentz theory could be met successfully, but it was almost immediately recognized that the whole model of a conducting metal had been oversimplified; the motion of the electrons could not be regarded as occuring in a force-free space. It was Bloch who in 1928 proposed a satisfactory electron theory of conduction on the basis of the wave mechanics.

If the electrons in a metal are regarded as wave systems rather than as material particles, it becomes possible to treat the whole problem of conduction in a novel manner. The electrons are still considered to be uncoupled, though the field in which any one electron moves is found by an averaging process over the other electrons. If the metal is at absolute zero, its lattice simply determines a periodic potential field for the electronic motions, and the electrical resistance by the immobile lattice is zero. An electron can move freely through a perfect crystal and a finite free path can only be due to imperfections in the lattice. In general the imperfections are caused predominantly by the thermal motion of the atoms and are strongly temperature dependent, increasing with increasing temperature. Impurities, however, also scatter the electrons, but in this case the free path will not vary appreciably with temperature. The resistance therefore consists of the "impurity resistance" and the resistance due to the thermal motion of the atoms. According to Bloch's analysis of the motion of an electron in a perfect lattice all the electrons in a metal can be considered to be "free", but it does not necessarily follow that they are all conduction electrons. Actually the free electrons in a solid form open and closed groups in much the same way as do electrons in an atom, and it is only when there are open groups that conduction electrons exist (Wilson 1931). In this way we arrive at a theory which can account for metals, semiconductors and insulators but not superconductors.

Thus, even though, in 1928 there was a successful theory of electrical conductivity, and some of the major problems of solid state physics were solved successfully, superconductivity *regarded as a phenomenon of infinite conductivity* was still not understood. Casimir puts forth two reasons for this:

> First of all because the phenomenology of superconductivity was not well established, although the phenomenon had been known for almost twenty years . . . one did not know what to explain, which made things more difficult. It was only after 1930 that one arrived at a more consistent picture of the macroscopic properties of superconductors. And, secondly, the theoretical explanation, as given about twenty years later, involved some further developments both of the mathematical formalism and also of its interpretation.[47]

Following Bloch's theory of metallic conduction, there were quite a few attempts to provide an explanation for superconductivity. We present the main ideas of Kronig (1933) and Dorfman (1933).

Kronig's model amounted practically to a re-introduction of Lindemann's hypothesis as regards only the very low temperatures: he assumed that below the transition point the conduction electrons formed a rigid lattice within, and independent of, the ionic lattice. This as a whole was fixed in position, but single chains of electrons were free to move along their length, and they did so under the influence of an external electric field. Under an external magnetic field the electron lattice would be perturbed and would gain energy so that the thermal agitation necessary to melt it, was reduced. Since the atomic lattice was not disturbed in the transition to the superconducting state, there should be no appreciable discontinuity in its non-electrical properties other than a possible small anomaly in the specific heat. In particular, there could be no sudden increase in thermal conductivity, due to the disappearance of resistance since below the transition point the electrons were incapable of exchanging energy with the ions.

Dorfman made no attempt to devise a definite model of a superconducting metal, but sought rather to derive the various experimentally observed phenomena from a single simplifying hypothesis of a mathematical nature. The assumption was that there were two distinct energy states possible for the conduction electrons, of which the lower corresponded to superconductivity and the higher to ordinary conduction accompanied by electrical resistance. It was like change of state affecting only the conduction electrons, the upper state being the Fermi electron-gas and the lower something analogous to a liquid or a solid. This hypothesis can be taken as being a consequence of several different metallic models such as, for example, the conception, which Dorfman ascribed to Einstein, of coupled electron swarms. This coupling of the conduction electrons was supposed to take place spontaneously.

Bloch himself tried unsuccessfully to solve the problem in 1928—29. His interpretation was suggested through analogy with ferromagnetism, where permanent magnetization had been explained by recognizing that parallel orientation of the magnetic moments of the atoms led to a lower energy than random orientation. Similarly, it seemed plausible to interpret current flow in a superconductor as the result of a correlation between the velocities of the conduction electrons that are energetically favoured and, therefore, manifests itself at sufficiently low temperatures. The picture of independent conduction electrons, which had otherwise been so fruitful in the theory of metals, did not provide for such a correlation.

> . . . I had already started to think about the problem and had realized that the explanation of persistent currents required a consideration of the previously neglected interaction between electrons. The idea independently expressed by Landau, was that it should thus be possible at low temperatures, to obtain a minimum of the free energy in a state of the metal with finite current. My own confidence in the idea was supported by the analogy with ferromagnetism whereby I saw a parallelism between permanent magnetization below the Curie point and persistent current below the critical temperature. I therefore started industriously to consider various types of interaction regardless of their possible origin, and to look whether the Schrödinger equation would allow stationary states of the electrons with non-vanishing current and a minimum of the energy. Once in a while I thought that I had indeed found such states but it never took Pauli long to point to some error in the calculations. While he did not object to my approach he became rather annoyed at my continued failure to come out with the desired answer to such a simple question. . . . Indeed I was so discouraged by my negative result that I saw no further way to progress and for a considerable time there was for me only the dubious satisfaction to see that others, without noticing it, kept on falling into the same trap. This brought me to the facetious statement that all theories of superconductivity can be disproved, later quoted in the more radical form of "Bloch's theorem": Superconductivity is impossible.[48]

The failure of such attempts to derive a quantum theory of superconductivity was mainly due to the same reasons which affected the development of thermodynamic arguments. Superconductivity continued to be regarded as a case of infinite conductivity. "In our opinion", F. London said,

> these attempts failed not only because the proper development of the quantum mechanics and quantum statistics of the electronic interactions in metals is in itself a formidable task but also *because the problem was not formulated quite correctly*. Indeed these attempts were defeated at the very start because they undertook to solve a problem which cannot be solved and which, fortunately, is not the one set by the phenomena.[49] (Emphasis added.)

3.3. Further experimental results and the phenomenological models

After 1930 there were more accurate data on the transition phenomena and a more consistent picture of the macroscopic properties of the superconductors emerged. In 1930 De Haas and Voogd found that the variation of the applied magnetic field and the superconducting temperature resembled curves of phase transition. Keesom and Clusius (1932), discovered the jump in the specific heat of liquid helium. A little later Keesom and Van den Ende (1932) found a somewhat similar jump at the critical temperature of tin. This result was subsequently confirmed by Keesom and Kok (1932).

Almost simultaneously a further property of the superconducting state was discovered. The thermal conductivity of a metal in this state was found to differ from that of the same metal at the same temperature, in which superconductivity had been destroyed by a magnetic field. The resistance—temperature curve, in the case of tin, forks at the transition point. Strangely enough it was the superconducting curve that appeared to be the continuation of the curve at higher temperatures. The curve with fields greater than the threshold value showed considerable discontinuity at the normal transition point (De Haas and Bremmer 1931). However, accurate experiments carried out shortly afterward on indium, showed that even without a magnetic field the thermal conductivity did indeed suffer a slight discontinuity at the normal transition point (De Haas and Bremmer 1932). Disappearance of resistance was no longer an isolated phenomenon, but was found to be accompanied by other changes and these experimental results seemed to make a thermodynamic treatment of the transition to the superconductive state feasible. Following the suggestion of Langevin,[50] the transition was described as a transition between two phases: the normal phase and the superconductive phase. A first trial in this direction had been made by Keesom, long before the jump in the specific heat was discovered. In 1924, because of Kamerlingh Onnes' illness, Keesom substituted for him at the Solvay conference where he had an interesting discussion with Bridgman on the thermodynamics of the disturbance by a magnetic field.[51] Bridgman sought to give an explanation of the threshold field by assuming that the magnetic permeability was different in the superconducting state than in the normal one. The transition could then be explained by a thermodynamical cycle. However, Lorentz and Keesom did not think that thermodynamics could be applied to this case. Keesom made the point that this disturbance was a typically irreversible phenomenon, since setting up an external magnetic field above the critical value $\mathbf{H}$ and thereupon reducing it to zero, was leading to a situation different from the original one. Nevertheless, writing down Clapeyron's equation, replacing P by the magnetic field $\mathbf{H}$ and V by the magnetic moment $\mathbf{M}$, he arrived at a relation between $d\mathbf{H}/dT$ along the transition line and the heat of transformation.[52]

In 1933, and after the discovery of the jump in the specific heat, Rutgers, starting from Ehrenfest's proposal about phase transition of the second order, derived a relation between the jump in the specific heat and the derivative of the magnetic threshold value with respect to the temperature.[53] The details, however, of the thermodynamic approach were worked out by Gorter (1933). His most interesting result was the derivation of the critical magnetic field:

> I obtained the field at which disturbance of superconductivity was thermodynamically possible, not entering into a discussion whether the mysterious irreversibility of the transition would considerably shift it to still higher fields. I suggested Rutgers join our efforts and that we write a paper together but he was of the opinion that the time was not ripe and so I published my calculation in the Archives of the Teyler's Foundation at Haarlem to which I then had moved.[54]

Another important contribution, which had a decisive influence both on the subsequent Leiden experiments and on the work of Meissner in Berlin came from von Laue in an attempt to discuss some experimental results of de Haas and Voogd. De Haas and Voogd[55] had found that, in the case of a tin wire made from a very pure single crystal, there was a marked difference in the behaviour of transverse and longitudinal fields. In a longitudinal field the normal resistance was restored suddenly at a definite field strength. In a transverse field, resistance began to appear when the measured field strength was only about half the corresponding longitudinal field. The resistance then increased gradually with the field at about the same field strength in both cases. Von Laue (1932) ascribed the result to the squeezing of the magnetic field lines which were supposed not to penetrate into the wire. To test this suggestion, De Haas (1933) studied the disturbance in a wire of ellipsoidal cross section. He kept the field constant and changed the temperature. His preliminary results were communicated at a colloquium organized by Debye in Leipzig in 1933. These can be summarised as follows: one gets essentially the same sort of behaviour, in a transverse field, whether one keeps the field constant and changes the temperature or cools the sample in a zero field and then switches on the field. The conclusion that one should draw from that seems obvious today: the magnetic-field distribution must be the same independent of whether the sample reaches the superconducting state in the presence or absence of an external field. Thus, there were no frozen-in fields. "But so strong was the doctrine of frozen-in fields that de Haas advocated another solution: Von Laue's theory is not good".[56] These developments forced gradually a change of attitude and "By the fall of 1933 we of the younger generation began to doubt the dogma of frozen-in fields and experiments were set up to settle this point".[57]

At that time there appeared a short letter to *Naturwissenschaften* by Meissner and Ochsenfeld (1933) which presented strong evidence that, contrary to every expectation and belief of the past twenty years, a superconductor expelled the magnetic field. Superconductors were found to be diamagnetic. The letter noted several experimental arrangements, involving either a pair of solid tin or lead cylinders or a cylindrical lead tube. In each case the sample was cooled below its transition point in a constant magnetic field. When the transition point was reached a sharp increase of flux was registered. They concluded that the magnetic flux in the specimen did not remain constant, but the lines of force were driven out of the superconductor, thereby increasing the flux in its neighbourhood.

Immediately after the discovery of the diamagnetic character of superconductors Gorter suggested that the zero-field inside the superconductors be a general characteristic of superconductivity, explaining, at the same time, that Tuyn's result was due to the formation of superconducting rings enclosing the normal state of matter. This meant that the condition $B = 0$ assumed in the previous thermodynamical treatment was not a restriction since superconductive states with $B \neq 0$ did not exist. In other words, a superconductor is a perfect diamagnet as well as a perfect conductor. Hence the phenomenon turned out to be reversible and thermodynamics could then, justifiably, be used. Casimir and Gorter (1934) discussed these considerations, extending them in various respects. They emphasised the idea of a phase transition and formulated the thermodynamics of this transition. Also some attempts were made at explaining the thermodynamics of superconductors in terms of a statistical model. Gorter and Casimir[58] proposed a two-fluid model in which the particle and current densities were expressed as the sum of normal and superfluid (superconducting) components:

$$n = n_n + n_s$$

$$J = n_s e u_s + n_n e u_n = J_s + J_n.$$

The component, n_s, which represented the superfluid condensate, was equal to the total electron density, n, at $T = 0$ °K, but decreased with increasing temperature and went to zero at T_c. The normal component, subject to the usual dissipation, came from electrons excited out of the condensate. This model gave a fairly good agreement with the early experimental results on the thermal properties (e.g. the T^3-dependence of the electronic specific heat in the superconducting state). The Gorter—Casimir model was set up so as to have the observed thermodynamic properties, but gave no information about the hydrodynamic or electrodynamic aspects of each fluid.

In 1935, Fritz and Heinz London extended the two-fluid model to give their phenomenological description of the electromagnetic properties of a superconductor. Fritz London was the first to draw attention to the fundamental implication of the Meissner—Ochsenfeld experiment: the property of perfect diamagnetism must be an intrinsic property of an ideal superconductor, and not merely a consequence of perfect conductivity. He concluded that superconductivity demanded an entirely new relation in which the current was connected not with the electric but with the magnetic field. The immediate reaction of everyone else to the Meissner effect was to try to fit it into Maxwell's electrodynamics but, with the permeability changing to zero, the equation became indeterminate.

The first such attempt to supplement Maxwell's equation was made by Becker, Heller and Sauter (1933). In fact they took up the argument that in a superconductor, or rather in a body without any resistance, you cannot have any change of magnetic field, and they pointed out that, because of the inertia of the electrons, an applied electric field would accelerate them steadily. Thus if there are n electrons per cm^3 of mass m, charge e and velocity v, we should have $m\dot{v} = eE$ and since the current density j is given by $j = nev$, we have

$$\mathbf{E} = \frac{4\pi\lambda^2}{c^2}\dot{\mathbf{j}}, \qquad \left(\text{we adopt the notation } \dot{x} \equiv \frac{\partial x}{\partial t}\right) \tag{2}$$

where λ is a material constant. Taking curls on both sides of (1) and using Faraday's law we find

$$\frac{4\pi\lambda^2}{c}\,\mathrm{curl}\,\dot{\mathbf{j}} = -\dot{\mathbf{H}}. \tag{3}$$

Substituting in Maxwell's equation curl $\mathbf{H} = (4\pi/c)\,\mathbf{j}$, we finally obtain

$$\lambda^2\nabla^2\dot{\mathbf{H}} = \dot{\mathbf{H}}. \tag{4}$$

Integrating with respect to time, (4) becomes

$$\lambda^2\nabla^2(\mathbf{H} - \mathbf{H}_0) = \mathbf{H} - \mathbf{H}_0 \tag{5}$$

where $\mathbf{H}_0$ is an arbitrary field (whatever field happened to be inside the body when it last lost its resistance). The general solution of (4), therefore, means that, practically, the original field persists in the superconductor for ever.

In our previous discussion of the magnetic properties of a perfect conductor the simpler result $\dot{\mathbf{H}} = 0$ (equation 1) was obtained instead of (4). The novelty of (4) is in showing that the value $\dot{\mathbf{H}} = 0$ (or $\mathbf{H} = \mathbf{H}_0$) is to be found only at a depth inside the metal greater than λ. Indeed, the solutions of this equation decrease exponentially as one recedes from the surface, where they are fitted into the values of the external field. This is a property of the type (4) differential equation.[59] There is no point in developing this form of the theory any further, for equation (3) merely leads to the equation $\mathbf{H} = \mathbf{H}_0$, with the modification that the magnetic field penetrates the body to a small but finite depth.

The conception of an accelerated current did, however, point the way to the correct equation in the macrosopic theory of superconductivity developed by Fritz and Heinz London. In 1935 the Londons proposed that the connection between magnetic field $\mathbf{H}$ and current density J_s for the pure superconductive case may be given by the equation

$$\frac{4\pi\lambda^2}{c} \operatorname{curl} \mathbf{j}_s = -\mathbf{H}. \tag{6}$$

Equation (6) can be obtained by time integration from (3) if it is assumed that the constant of integration is zero ($\mathbf{H}_0 = 0$) and it was considered as a completion of Becker, Heller and Sauter's formalism by fixing the integration constant of the magnetic field according to the Meissner effect.

Equation (6) leads immediately to

$$\lambda^2 \nabla^2 \mathbf{H} = \mathbf{H}. \tag{7}$$

For large specimens, the characteristic feature of the solutions of this equation is that they decay exponentially into the interior of the specimen. In a distance λ from the surface the field is practically zero.[60] Here Meissner's experimental result is representable by equation (6) with one restriction, namely that the magnetic flux decreases, not abruptly on the surface, but continuously in a very small interval below the surface. Equations (2) and (6) provide an adequate description of a macroscopic superconductor; they describe the zero resistance and the Meissner effect respectively.

F. London suggested that a supercurrent is to be regarded as a kind of diamagnetic current. Moreover, Fritz and Heinz London supposed "the electrons to be coupled by some form of interaction. Then the lowest state of the electron may be separated by a finite distance from the excited ones".[61] This may be the earliest suggestion of an energy gap.[62] In a discussion at the Royal Society, F. London (1935) went further and proposed that superconductivity is the result of a "rigidity" in the ground state electronic wave function that gives rise to diamagnetic currents maintained by a magnetic field, thus suggesting a way of reconciling superconductivity with quantum theory.

3.4. Towards a microscopic theory

That quantum theory was essential for explaining superconductivity was strongly suggested from a theorem stating that a classical system can exhibit no diamagnetism. Bloch and Brillouin[63] proved that the most stable configuration of electrons, in the absence of an external field, will, in all probability, display no current.But, in a diamagnetic atom we have an example of a permanent current flowing in a system which is in its most stable state. The apparent contradiction between London's theory and Bloch's theorem was avoided by the fact that when there is no external magnetic field, Bloch's theorem is no longer valid.

In attempting to "sketch the programme which seems to be set by our equations to a future microscopical analysis",[64] F. London regarded "the total supraconductor . . . as a single big diamagnetic atom".[65] For atomic or molecular systems, the susceptibility is given by

$$x = \frac{-Ne^2}{6mc^2} \sum_i \overline{r_i^2},$$

where N is the number of atoms per cm^3, and r^2 is the mean square radius of the orbit of the electrons. A perfect diamagnetism corresponded to $x = 1/4\pi$ giving $\mathbf{B} = \mathbf{H}(1 + 4\pi x) = 0$. A value of this order required large orbits for the electrons. But, a model with a large effective r and large diamagnetism did not necessarily have all the properties of a superconductor, such as, for example, persistent currents. London continued as follows: He took the general expression for the electric current density to be

$$J = \frac{he}{4\pi im}(\psi \text{ grad } \psi^* - \psi^* \text{ grad } \psi) - \frac{e^2}{mc^2}\psi\psi^* A,$$

where ψ is the wave function of a single electron in the self-consistent field of the others. Then $\psi\psi^*$ gives the value of the statistical expectation for this electron at every point of the space. Summing over all electrons, $\Sigma\psi\psi^*$ gives the number n of electrons per cm^3. The magnetic field is described by the vector potential **A**. In the absence of a magnetic field ($\mathbf{A} = 0$) $\psi = \psi_0$ and the current density vanishes. But, in a normal metal, the wave function responds to the magnetic field in such a way so that there is cancellation between the "paramagnetic" contribution involving the gradient terms and the diamagnetic contribution proportional to the vector potential, **A**, leaving only a very weak diamagnetism. London proposed that a superconductor differs in that there is a rigidity of the wave function such that it is essentially unmodified by the magnetic field. If the wave function is unchanged, the paramagnetic contribution (i.e. the terms in brackets) vanishes even in the presence of the field. This leaves the diamagnetic term which gives an equation similar to Londons' phenomenological equation:[66]

$$\mathbf{J} = -\frac{ne^2}{mc}\mathbf{A}.$$

To understand the significance of this result one must have in mind that in a normal metal the average value of the momentum operator, P^{ave}, changes in a magnetic field in such a way that the velocity is practically zero. F. London suggested that the reason that P^{ave} does not change in a superconductor when the field is applied, is that there is a long range order which maintains the local average value of the momentum constant over large distances in space. This order would be maintained even in the presence of the magnetic field. The ordered ground state was regarded as a single quantum state extending throughout the metal.

Such an interpretation characterises the superconductor as "a quantum mechanism of macroscopic scale" which requires "a kind of solidification of condensation of the average momentum distribution".[67] We now know that these ideas were essentially correct and they were to form the basis of one of the most remarkable conceptual developments in the history of quantum mechanics. An interesting prediction of the London theory was the quantization of flux in a superconducting ring in units of hc/e. This prediction was confirmed many years later (after, in fact, the proposal of successful microscopic theory), and considered as a "demonstration of the width of London's vision".[68]

Undoubtedly, the London theory defined the goals for the microscopic theory of superconductivity, by emphasising the diamagnetic approach, the long-range order in momentum space, the rigidity of the wave function in the superconducting state and the relation between the current density and the vector potential. Nevertheless, detailed comparison between the London theory and experimental results showed several discrepancies. These discrepancies stimulated the developments of models by Pippard and by Ginzburg and Landau.

The London equation is a local equation because it relates the current density at a point r to the vector potential at the same point. Pippard (1953), on empirical grounds, proposed a generalization of the London equations in such a way as to give a nonlocal relation between current density and magnetic field. The basis of his theory was his concept of coherence: that the range of order or the wave functions of the condensed superconducting phase extended over rather large regions of space, of the order of 10^{-4} cm in a pure material, and the coherence length ζ was a measure of nonlocal effects. The Pippard theory, although it seemed to be merely a modification of the London theory, represented a significant change from the earlier theory in its treatment of the problem of how an applied field influenced the superconducting electrons. A basic point of the London theory was the absolute rigidity of the superconducting wave function in the presence of a field. Pippard abandoned this point and suggested instead that a perturbing force acting at one point in the superconductor would be felt over a distance ζ, and, conversely, the response at a point due to a spatially extended perturbation would be obtained by integration over a finite region surrounding the point. Pippard arrived at the concept of the range of coherence partly empirically and partly on the basis of his interpretation of the nature of the superconducting state. His coherence length was the shortest distance in which a significant change of electronic structure in the superconductor could occur.

Meanwhile Ginzburg and Landau[69] in 1950 developed the thermodynamics of a model in which the energy needed to produce a change in the superconducting state over any distance was explicitly included in the theory. Ginzburg and

Landau noted certain inadequacies of the London theory and defined a parameter ω which was a measure of the order in the superconducting phase, so that it was zero for $T > T_c$ and increasef smoothly as T was reduced below T_c in zero field. They identified ω with the square of an effective wave function ψ, defined so that ψ^2 is equal to the concentration of superconducting electrons. ψ did not describe a single particle, but, rather, the motion of the superfluid condensate as a whole. The Ginzburg—Landau theory predicted the right dependence of critical field upon T (at least close to T_c) and provided a justification for the very large positive energy needed to explain the Meissner effect. The Ginzburg—Landau theory leads to the usual London theory when the effective wave function is a constant. Indeed it is a "local" theory and distinct from the nonlocal theory of Pippard. Moreover, considerations based on the microscopic theory indicate that the Ginzburg—Landau theory is strictly valid only near T_c, but it did help to give a good qualitative understanding of superconductivity at all temperatures.

Summarizing the features that a satisfactory microscopic theory of superconductivity was required to pocess, one had the following: the superconducting state should display a unique diamagnetic relation between the current and magnetic field (London); an energy gap has to be present in the excitation spectrum of the electrons; the current-field relation is nonlocal, with a range of coherence 10^{-4} cm (Pippard); the theory should involve an "order parameter", which went smoothly to zero as $T \rightarrow T_c$ (Ginzburg—Landau).

A major obstacle to progress toward a microscopic theory was the ignorance of the effects of interactions, and how to treat them. It was not really understood why the independent electron model of metals worked so well. It was not known what part of the interaction between the electrons was "used up" in the normal state and what remained to bring about the phase transition. The plan for a successful microscopic theory was to identify first that part of the electron—electron interaction responsible for superconductivity.

One of the first definite proposals for such an interaction was due to Heinsenberg.[70] In 1947 he suggested that the singular part of the Coulomb interaction (because of its long range) could lead to superconductivity, and characterized superconductivity as a state of electronic lattice order. From Heisenberg's model one would infer that a state with spontaneous currents should be stable even in the absence of any applied magnetic field. Apparently the idea originated from an analogy with ferromagnetism, as the early attempt of F. Bloch, and ran into the same difficulty: there is no "hysteresis" so long as one stays within the limits of the pure superconducting state. Heisenberg's model was, on this account, severely criticized by F. London (1948, 1949). London accepted the Coulomb interaction of the electrons as essential for the establishment of the superconducting state; the "essential difference . . . is the assumption that the perfect conductivity rather than the diamagnetism is the primary feature of the phenomenon". Nevertheless, Koppe (1950) proposed a two-fluid model, based loosely on the Heisenberg theory of superconductivity and Goodman found good agreement with Koppe's model.[71]

Another interaction which was suggested as being responsible for superconductivity was the magnetic interaction between electrons.Welker (1938) tried, without success, to base a theory on the magnetic exchange interaction between particles. These and other similar attempts to give a microscopic theory of superconductivity by including interactions omitted from the Bloch theory of conduction, failed almost completely.[72]

There is one more interaction, however, the electron—phonon interaction. But, since the discovery of superconductivity there had been a widely and firmly held belief that the ion masses, being so much larger than the electron masses, could not play an important role in the establishment of the superconductive state. Fröhlich in 1950 "conceived the idea that just the opposite of the 'dictum' contains the truth".[73] The development of such an approach was closely connected with the introduction of new methods, those of field theory, into solid state physics. Methods to treat interactions between particles were first devised for quantum field theory in the late 1940's by Fröhlich, Pelzer and Zienau (1950), for a single electron in polar crystals — a problem much simpler than superconductivity. The field theoretical treatment showed that the kinetic energy of the ions attached to a moving electron may be much smaller than the kinetic energy of the electron.

> Clearly such a result was of great interest in connection with the "dictum" that ions, in view of their heavy mass, should be of no importance for superconductivity and the application of appropriate field theoretical methods to metals was called for.[74]

Fröhlich, then, applied field theoretical methods to the interaction of the electrons in a metal with the lattice vibrations, and he concluded that the interaction would lead to an attraction between the electrons.

> In fact prior to my introduction of phonon-induced electron interaction it seemed ridiculous to assume attractive forces between electrons, which is probably the reason why the theoretical treatment of superconductivity was held up for so long.[75]

At the same time, and independent of what was implied by these theoretical developments, experiments were undertaken to determine whether or not there was a dependence of critical temperature on isotopic mass.[76] These experiments showed, surprisingly at the time, that T_c varied inversely with the square root of the isotopic mass. Thus the mass became an important parameter when the motion of the ions was involved, and this, in turn, suggested that superconductivity could be derived from some sort of interaction between the electrons and zero-point vibrations of the lattice.

> It must be remembered that the isotope effect was derived as as consequence of the theory before it was discovered although the first experiments were performed before publication of the theory. To most physicists it came as a great surprise; in fact I know of one laboratory which earlier had rejected an offer of isotopes for low temperature investigation. For a theoretician it provided a welcome confirmation of the basic idea. The main reason for accepting it was in the first place, however, the fact that an attractive interaction had been derived as a direct consequence of the semiempirical free electron model as applied by F. Bloch to the theory of normal electric resistivity.[77]

Fröhlich's 1950 paper was followed by Bardeen's attempt to base a theory of superconductivity on the interaction between electrons and phonons. The model he used turned out to be very similar to that used by Fröhlich.[78] However models of this type were essentially independent electron models and could not explain the cooperative features of the superconductive state already implied by London's theory. Furthermore, the predicted condensation energy was too large. The use of perturbation theory in a situation where the properties of the state were drastically modified was unjustified and, as Schafroth showed, the theory could not lead to the Meissner effect.[79] Yet neither Schafroth, nor most of the other researchers doubted the correctness of the basic assumption that electron-phonon interaction should be principally responsible for superconductivity.[80] To the features which a microscopic theory of superconductivity was required to have, one had now to add that of the electron-phonon interaction.

> Thus, the outstanding problem was to see how an energy gap might follow from a microscopic theory based on electron and phonon interactions as implied by the isotope effect.[81]

The following figure (Figure 4) illustrates the status of the theory by 1955

The first step away from perturbation theory was the use of a canonical transformation by Fröhlich (1952) to remove the electron—phonon interaction and to replace it by an effective electron—electron interaction, though his model did not explicitly[82] take the Coulomb repulsion into account. There remained the question of whether the much larger Coulomb interaction would overwhelm the Fröhlich interaction, but Bardeen and Pines (1955) showed that Coulomb interactions did not change the essential features of Fröhlich's result for the phonon-induced interaction.[83] Moreover, for pairs of electrons whose energies were within a characteristic phonon energy of the Fermi surface, this attractive

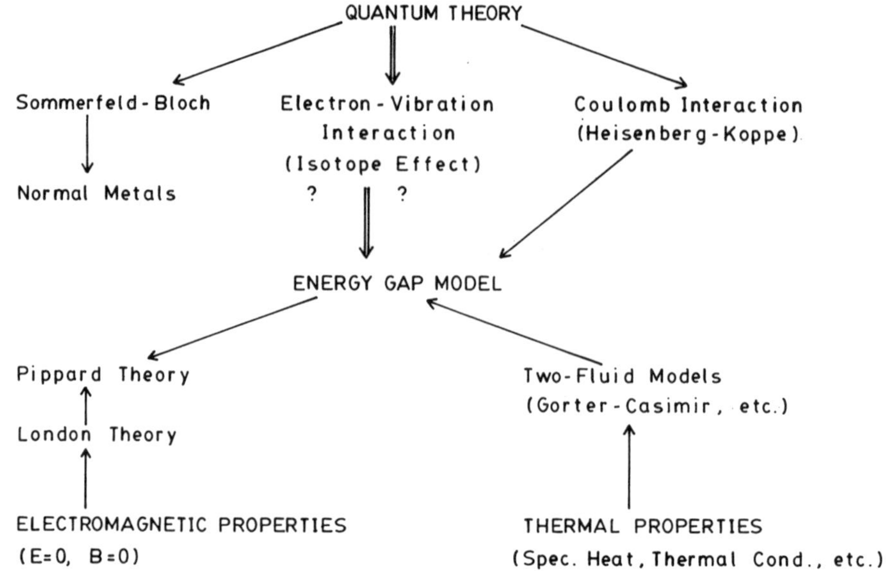

Fig. 4.

interaction could dominate the repulsive screened Coulomb interaction. This essentially became the criterion for superconductivity.

At about the same time Ginsburg and Schafroth (1954, 1955), put forth the idea was that, if electrons are associated in pairs, the pairs would obey Bose–Einstein statistics, and that superconductivity, like superfluidity in helium, would be a consequence of Bose condensation.[84] With Blatt and Butler, Schafroth (1957) tried to prove that a Fermi gas with appropriate interaction between the particles would behave like a gas of charged bosons. Their idea was to take into account two-particle correlations between the electrons and to ignore higher correlations so that the system would behave like a gas of Bose molecules whose members could be dissociated into two fermions.

> The most dramatic moment I remember having to do with superconductivity came at the end of Feynman's talk at the Seattle theoretical physics conference in September 1956, when he announced that, although he had solved the problem of superfluidity, he had spent many months computing on the problem of superconductivity and had failed utterly. Some radical new idea was needed to solve the problem, he said. At his point John Blatt leaped up on the stage and anounced, "We have the idea, and we have solved the problem". The idea was pairs, indeed, but the Australian group was very far from a formal solution of the problem.[85]

Although Blatt, Butler and Schafroth were unable to calculate the properties of the system from their model, "this was probably the first realistic attempt to go beyond a one electron picture of a superconductor to explain the cooperative properties".[86]

The next major step was made by Cooper (1956), who showed that if there is an effective attractive interaction, a pair of quasi-particles above the Fermi sea will form a bound state no matter how weak the interaction. His calculations showed definitely that, in the presence of attractive interactions, the Fermi sea which describes the ground state of the normal metal is unstable against the formation of such bound pairs. Cooper also pointed out that it was not correct to think of the pairs as discrete entities and, thus, a picture of Bose–Einstein condensation of isolated pairs was not a possible one. He, moreover, suggested that, to the extent that the many-body system could be regarded as a collection of such bound pair states, it would display many of the properties that Bardeen had argued, in a review article which appeared in 1956, were required to explain superconductivity.[87] "A true many-body approach was required to find the ground state wave function of a superconductor".[88]

Bardeen, Cooper and Schrieffer (1957a) collaborated closely in working out the decisive step to the theory of superconductivity. In a preliminary note, submitted in February and published in April 1957, they showed how to generalize the Cooper pair states to the many body problem at absolute zero. This letter included an expression for the ground state energy difference between normal and superconducting states and for the energy gap at $t = 0$ °K. In the spring of 1957 they extended their theory to obtain the excitation spectrum and made detailed calculations for various thermal and transport properties at temperatures above absolute zero. They worked out an expression for the specific heat and showed that the transition at T_c is of second order with a jump in specific heat but not latent heat. For simplicity, their calculation of thermal, electromagnetic and other properties of superconductors were based on a simplified model which involved three parameters: the normal density of states in energy, the velocity of the electron at the Fermi surface and an interaction constant which gave a measure of the phonon-induced attraction between electrons. Only the latter involved the superconducting state and could be related to the transition temperature, but the parameters could be combined to give two lengths: the London penetration depth and the Pippard coherence length. The full paper by Bardeen, Cooper and Schrieffer (1957b) appeared the fall of the same year. Their "model is exactly of the sort which should account for superconductivity according to London's ideas".[89] Indeed, in the ground state, electrons were associated in pairs of opposite spin and momentum or, when there was current flow, in pairs of exactly the same momentum. The common momentum of the paired states gave rise to a "a kind of solidification or condensation of the average momentum distribution" as suggested earlier by F. London, to account for his tentative proposal that a superconductor is indeed "a quantum structure on a macroscopic scale".

Thus, forty six years after the first experimental observation of superconductivity, a microscopic theory which explained most of the properties of superconductors was finally developed. Its agreement with experiment was impressive.[90]

> If there was a discrepancy, it was usually found, on rechecking, that an error had been made in the calculation. All of the hitherto puzzling features of superconductors fitted neatly together like the pieces of a jigsaw puzzle.[91]

CHAPTER 4

Superfluidity: old concepts in search of new contexts

4.1. Early experimental investigations

Helium "has always been a most remarkable and entertaining substance. Consider the manner of its discovery. Most of the rare elements have been found by painstaking search and careful chemical isolation, but helium was discovered almost by accident, not on the earth but in the sun! In fact, after the first discovery of helium in the solar atmosphere, nearly thirty years were to elapse before it was found to be present on the earth".[1] There was an immediate race to separate the gas in pure form in quantity and to try to liquefy it.[2] Thirteen years of unsuccessful attempts (Olszewski in 1896, Dewar in 1901, Travers in 1903 and Olszewski again in 1905) were to elapse before it was liquefied by Kamerlingh Onnes[3] on 10th July 1908.[4]

As we have already shown, the liquefaction of helium was first and foremost a triumph of the systematic researches of Kamerlingh Onnes particularly on the work started by van der Waals, complemented by the progress in cryogenic techniques. Helium has a unique position in low-temperature physics; it is indispensable for the production and maintenance of low temperature and itself has most peculiar properties. The temperature at which helium liquefies (4.2 °K at atmospheric pressure) is lower than that of any other gas and, unlike all other substances, it does not solidify under its own pressure, remaining liquid down to absolute zero, even though it is possible to convert liquid helium into a solid under increased pressure. At temperatures not too far below 4.2 °K helium behaves like an ordinary liquid. But below 2.19 °K — the so-called lambda point — helium, although remaining liquid, undergoes another type of transition; a transition which is marked by the abrupt change in the values of many thermodynamic parameters such as density, specific heat of vaporisation etc.

It has been claimed that, on the day Kamerlingh Onnes liquefied helium, he may have observed a dramatic change in the various properties of the liquid helium but there does not appear to be any evidence for this.[5] Trying to reach the triple point he had pumped away the vapour to such a low pressure that he obtained a temperature much lower than 2.19 °K.[6] He had made a rough determination of the liquid density and found it to be about 0.15, an astonishingly low value.[7] This, and particularly the density maximum he discovered a little above 2 °K in 1911,[8] led him, when receiving the Nobel prize in 1913, to the suggestion that "such an extreme could possibly be connected with the quantum theory".[9] He was also prompted to conclude his report to the Third International Congress of Refrigeration by proposing "further experimental study of helium . . . including the investigation of related properties such as viscosity, capillarity, specific heat, refraction index and dielectric constant . . .".[10] Some of these investigations were undertaken in the subsequent years by himself and his collaborators; but many of the properties of liquid helium below 2.19 °K were actually discovered almost twenty-five years later. Many of these results were, in fact, reported in the remarkable volume of *Nature* No. 141 in 1938.

In retrospect it seems quite surprising that it took so many years, from the observation of the density maximum of liquid helium to the discovery of all those properties which eventually characterised the superfluid state of helium. According to Casimir "this may be connected with the views of Kamerlingh Onnes who emphasised quantitative measurements rather than qualitative observations"[11] . . . "And that is not always the best way to discover new effects".[12] It is worth remembering that low-temperature work was faced with considerable technical difficulties.[13] In order to carry out experiments not only large quantities of liquid air were required, it was also necessary to have hydrogen and helium liquefiers equipped with the necessary compressors, pumps and gasometers, and considerable staff of trained mechanics. Cryogenic work at liquid helium temperatures was coupled with even more intricate technical problems. Even the supply of helium gas was at that time a source of difficulties, and the cost of helium gas with which Kamerlingh Onnes first produced the liquid has been estimated at $2500 per cubic foot.[14]

More than twenty five years elapsed between the liquefaction of helium and the discovery of all those properties of helium which comprise the phenomenon of superfluidity. It is interesting to notice that the paper in which Kamerlingh Onnes described the discovery of the density maximum, had a second part on the resistance of platinum at helium temperatures; an important step towards the discovery of superconductivity. It is probable, then, that "the interruption of dramatic progress in the field was caused in large measure by the discovery in 1911, also by Onnes, of superconductivity in metals".[15]

The investigations carried out on the two forms of liquid helium (Helium I and Helium II below 2.19 °K) may be divided, into four categories:[16]

(1) "Routine" measurements which were intended, at first, to find how the liquid acts as absolute zero is approached, and were later directed to the discovery of discontinuities at the lambda point. Under this heading one can include:

vapour pressure, density, and coefficient of expansion, compressibility and the velocity of sound.

(2) Measurements of the thermal properties, *viz.*, the specific heats of the two forms of liquid helium, the possible latent heat accompanying the transformation, and the latent heats of solidification and vaporisation.

(3) Investigations of various physical properties which might throw some light on the molecular structure of the liquid: determinations of the surface tension, dielectric constant, refractive index, light scattering and Raman effect, electrical resistivity, and the structure as indicated by the use of X-rays.

(4) Measurements of the viscosity and thermal conductivity, and all experiments with helium below 2.19 °K in a state of flow. These are the properties which distinguish helium from all other liquids.

In his experiments to reach very low temperatures with liquid helium Kamerlingh Onnes worked with an arrangement that consisted of a small Dewar vessel in a large one filled with liquid helium. Quite accidentally be observed a transfer, at a striking speed, of helium from one vessel to the other, until the levels in the two containers had reached the same height. He proposed that his observation might be explained by assuming a distillation from one vessel to the other. However, the transfer was actually much too fast to be explained by an ordinary distillation. He briefly described this observation in a report to the Faraday society in 1922 but no further notice was taken of this remarkable behaviour of helium:

> The speed of readjustment by this distillation was striking. A correct judgement on this phenomenon will only become possible, when the determinations we have in view concerning the heat conductivity of glass, of helium vapour and of liquid helium have been carried out, whilst a knowledge of the latent heat of evaporation and of the specific heat of liquid helium and of glass and of the viscosity of gaseous helium is also desirable.[17]

Kamerlingh Onnes' characteristic hesitation to draw any conclusions, before completing an exhaustive series of experiments and measurements, prevented him foretelling the existence of the phenomenon of a creeping film, rediscovered in 1938. Neither Kamerlingh Onnes nor anyone else at that time had recognised the significance of such an observation. Certainly Mendelssohn did not suspect any connection with it ten years later when his "calorimetric measurements were spoiled by a layer of helium which had evidently formed on solid surfaces".[18] In 1936 Rollin had suggested that his failure to reach very low temperatures (by pumping off, like Onnes, the vapour above liquid helium) might be due to the high thermal conductivity of a helium film covering the inside wall of his apparatus, and this erroneous explanation was also restated by Kikoin and

Lasarew in 1938. It was Daunt and Mendelssohn who eventually discovered and published in the above mentioned volume of *Nature* the remarkable phenomenon of the "supersurface" or "creeping" film.

In 1924, Kamerlingh Onnes and Boks made more elaborate measurements of liquid helium density and found that the density—temperature function had a "sharp maximum with a discontinuity of its slope" at that temperature.[19] Care was then taken to eliminate errors due to a possible anomaly in the expansion of the glass vessel. But despite the fact that Kamerlingh Onnes thought the result was important, its significance was not appreciated at the time. The analogy with the density maximum in water was tempting enough to exclude other explanations. Nevertheless, the realization that there was a lot more to be learned about helium did originate from these observations.

In the following two years, Dana, Van Urk and Keesom[20] investigated the surface tension, the heat of vaporisation and the specific heat of liquid helium. The results of these measurements clearly pointed to "a jump in the value of the surface tension and in the heat of vaporisation", and possibly should indicate "that near the maximum density *something happens to the helium*, which within a small temperature range takes place perhaps even discontinuously".[21] It was in the last year of his life[22] that Kamerlingh Onnes reported the suspected existence of two states of liquid helium, since "they had found values of the specific heat which were so high that they did not dare to publish them. They felt sure that something had gone wrong with their measurement equipment".

In 1927 Keesom and Wolfke, summarizing results of various experiments on the liquid helium properties and their own measurements of the dielectric constant of liquid helium between the boiling point and 1.9 °K,[23] concluded that:

> we have to do here with two different states of liquid helium, which transform into each other at the temperature corresponding to the pressure mentioned. Of those two phases the liquid helium II (stable at the lower temperature) compared with helium I has: a smaller density, a greater heat of varporization, a smaller surface tension, while the transformation liquid helium II — liquid helium I takes place with an absorption of heat . . .[24]

It was the first time that the terminology about the two kinds of helium was introduced.

In 1930 Keesom and Van der Ende observed quite accidentally that liquid helium II passed with remarkable ease through certain extremely small leaks, which at a higher temperature were perfectly tight for liquid helium I and even for gaseous helium. This observation seemed to indicate an enormous drop of the viscosity when liquid helium was cooled below 2.19 °K. However, measurements of the viscosity of liquid helium by Wilhelm, Misener and Clark in 1935 in Toronto and by Keesom and MacWood in 1938 in Leiden[25] showed that the viscosity below 2.19 °K, although an appreciable drop was found, varies continuously and is certainly not very different from the viscosity of helium I. In both investigations oscillating cylinders or disks were used which precluded the observation of superfluidity, thus leaving another remarkable property to be discovered a little later, by researchers from Cambridge and Moscow, where the viscosity of helium, measured by letting it flow through narrow capillaries, was found to be smaller by a factor of million.

The experimental fact that the viscocity determined by a rotating disk method is *so much different* from that derived from flow through capillaries is in itself a most impressive deviation from classical hydrodynamics. But, the "technicians in the Leiden Lab were not surprised. They had noticed that you can pump against a leak in a vessel immersed in liquid helium as long as the temperature is above 2.19 °K; at lower temperatures it is hopeless. And again nobody had paid attention to this fact".[26]

The first description of the change in visual appearance that accompanies the transformation from liquid helium I to liquid helium II was published by McLennan, Smith and Wilhelm in 1932. Describing the boiling of liquid helium under reduced pressure, as its temperature fell through the region of the density maximum discovered by Kamerlingh Onnes, they noted that:

> the appearance of the liquid underwent a marked change, and the rapid ebullition ceased instantly. The liquid became very quiet and the curvature at the edge of the meniscus appeared to be almost negligible.[27]

That same year Keesom and Clusius (1932) reported that the specific heat curve had an "extremely sharp maximum", although there was no latent heat for the transition from helium I to helium II, and then confessed ignorance as to the "inner causes" of the transition. These investigations on the specific heat of liquid helium "gave rise to some questions which made a nearer [sic] investigation of the anomaly of the specific heat of the substance . . . desirable".[28] With his daughter Anna Petronella, Keesom repeated the specific heat measurements with greater accuracy. On passing through 2.19 °K they noticed that their data underwent a sudden change.[29] Their paper contains the first public proposal of the now-established name for the lambda point — the point where we have the helium I—helium II transition:

> Leaving open the question whether the fall of the specific heat really occurs discontinuously in the strict sense of the word, or in a very small temperature interval, it will be appropriate from an experimental point of view to consider the fall to occur abruptly. For convenience sake it is desirable to introduce a name for the point at which this jump occurs. According to a suggestion made by Prof. Ehrenfest we propose to call that point, considering the resemblance of the specific heat curve with the Greek letter λ, the *lambda-point*.[30]

During another repetition of the specific heat measurements, the Keesoms[31] advanced the assumption that at the lambda-point the heat conductivity must rise strongly because the observed temperature drifts changed suddenly.[32] However, the connection with the remarks of McLennan and his colleagues was not recognized.

The next step was a proper determination of the heat conduction of helium II. In 1936 they tried to measure it by a conventional method and then, finally, realised that below the lambda-point the "heat conductivity is about 200 times that of copper at ordinary temperatures, or about 14 times that of very pure copper at liquid hydrogen temperatures. Hence liquid helium II is by far the best heat conducting substance we know".[33] A more accurate method was adopted in 1937 by Allen, Peierls and Uddin. They found, just as Keesom and his daughter had, that the heat conductivity of liquid helium reached enormous values below the λ-point which decreased as the temperature was lowered. Even more astonishing, however, was their observation that the thermal conductivity increased at all temperatures with diminishing heat current. The results were most remarkable indeed: there was an almost infinite thermal conductivity which, did not follow classical rules. The concept of "heat conductivity" in the accepted sense as a constant ratio of heat current density to temperature gradient had thus lost its usefulness when dealing with liquid helium II, and Allen, Peierls and Uddin ended their letter to *Nature* by saying that they could offer no explanation for these results "for which there is no analogy in the behaviour of other substances".[34]

The establishment of the anomalously high thermal conductivity in helium II was decisive for the boost of the research activities about the properties of liquid helium.

4.2. "Unnatural" phenomena in *Nature*: six letters to the editor

First Letter. As we have already noted the first investigation which led to a new discovery had quite a modest aim. It was to repeat the Leiden heat conductivity measurements with greater accuracy. Using a simple but very sensitive apparatus the Cambridge team not only confirmed the high heat transport found in Leiden but noticed that the heat current in liquid helium II is not proportional to the temperature gradient.[35] This indicated that the heat transfer in liquid helium II cannot be characterized simply by large value of the ordinary heat conductivity coefficient, and that accordingly the customary theory of heat conduction had to be replaced by something else. The term "*thermal superconductivity*"[36] was then used to characterize this phenomenon.

Second and Third Letters. Six months later, in the January 1938 issue of *Nature* there appeared two communications. One was by Kapitza (1938) in Moscow and the other by Allen and Misener (1938) in Cambridge. They both reported viscosity measurements of liquid helium II by flow through a narrow slit or capillaries.

Both yielded the same result: in contrast to the results of Keesom and MacWood (1938), and confirming an observation of Keesom and Van der Ende (1930), the viscosity of helium, at least when measured by the flow through narrow slits, below the λ-point fell to an immeasurably small value and, moreover, it was not proportional to the pressure gradient. The flow velocity increased, becoming less pressure-dependent for the narrow flow channels and Kapitza, for the first time, used the term "*superfluidity*". These experiments did not indicate the presence of a viscous, laminar, or turbulent, flow and they could not be discussed on the basis of classical hydrodynamics. The failure of classical theory became particularly striking in view of the results of Keesom and MacWood, which at first sight seemed to contradict the observations of Kapitza, Allen and Misener. Here, then, was a dilemma: "two time tested and, we thought, reliable methods of measuring viscosity giving values which were poles apart".[37] As the various experiments to measure the viscosity of liquid helium were performed in Leiden, Toronto, Moscow and Cambridge, "It seemed as if the value depended not only on the type of the experiment but also on the place where the experiment had been performed; a state of affairs which is scientifically undesirable".[38]

Fourth Letter. One month later the third of the new discoveries "perhaps the strangest of all" was reported by Allen and Jones and published in the February 1938 issue of *Nature*. The discovery was made almost by accident when Allen and Jones, interested in extending the heat conductivity measurements to lower and lower temperature differences, used an apparatus consisting of a reservoir and a smaller vessel, both filled with liquid helium and connected by a fine capillary. When they supplied heat to the inner vessel, they observed "the almost unbelievable effect" that the inner helium level, far from being depressed, seemed to rise above that of the reservoir. The rise increased with heat input and, for constant input, with falling temperature. This was the "thermomechanical effect", a mass flow of helium opposing the heat current. In one of their experiments Allen and Jones used a powder-filled bulb, open at the bottom and with a narrow orifice at the top. When they heated the powder by shining a light on it, they observed a jet of liquid helium rising from the upper end to a height of several centimeters. The name "*fountain effect*" was given to this phenomenon.

The more appropriate term of "thermomechanical effect" indicated that one dealt with a motion of mass produced by heat and suggested a heat engine whose reversibility was demonstrated by Daunt and Mendelssohn. Here a small Dewar vessel is open at the bottom end and contains, at this point, a plug of fine emery powder. The vessel also contains a very sensitive thermometer. The vessel may be lowered into a bath of He II or, alternatively withdrawn. Thus liquid flows either into the bulb or out of it. When liquid flows into the bulb, a decrease in the temperature is observed on the thermometer. When liquid flows out of the vessel, an increase in temperature of the same amount occurs. It was, then, inferred that sources of heat in helium II produce motion of the liquid and, conversely, a motion of the liquid produced temperature differences.

The "fountain effect" demonstrated the *inadequacy of the concepts provided by the dominant theroretical framework.* Extremely small temperature differences between reservoir and inner vessel were sufficient to produce a very large convection. It seemed, thus, impossible to treat the hydrodynamical and calorific properties of liquid helium independently.

Fifth and Sixth Letters. A few months later another new phenomenon was added to the list, that of the mobile, "*creeping*" or "*supersurface*" film. Daunt and Mendelssohn (1938) in Oxford, and Kikoin and Lazarew (1938) in Kharkov demonstrated by direct experiments that helium flowed from one container to another inside it (or outside it depending on the relative height of the liquid helium surface) by means of a film 3×10^{-6} cm (that is, of the order of 100 atoms) thick, formed on the walls. The liquid in the film may travel at a velocity of more than 30 cm s^{-1}.

The propensity to form a film is, in itself, not a unique property of liquid helium II. Such films are formed by any liquid which wets a solid surface, but the viscosity of an ordinary liquid is such that the film forms slowly and moves scarcely at all. Helium II is the only fluid which, owing to its superfluidity, forms a swiftly moving film.[39] "The most important feature of our experiments was that the helium film, being so much thinner than Kapitza's or Allen's capillaries, provided the pattern of superfluidity *par excellence*, including the ideal explanation of the thermomechanical effect".[40]

With the discovery of film transfer we come to the end of an era during which the various properties of liquid helium were studied, and its most peculiar behaviour established. This prompted physicists to consider helium II as the only representative of a "fourth state" of matter. The community of physicists was confronted with a new phenomenon, and it was evident that its explanation would require radically new approaches.

4.3. A new kind of "order"

Towards the end of the 19th century atoms were still regarded as being more or less indivisible and with no inner structure, and physicists were mainly concerned with finding out how these atoms were arranged to form molecules and how they were grouped in different states of aggregation "without bothering much about the nature of the forces which were responsible".[41] Moreover it was thought that by decreasing the temperature one would be able to liquefy or solidify even gases with very low heats of evaporation or fusion, and that nothing "unexpected" would happen.

The whole outlook was changed by the development of quantum theory and the discovery and interpretation of the quantum effects in low temperatures. The first indication of such effects was, in fact, provided by liquid hydrogen which Dewar in 1898 found to evaporate a good deal faster than the excellence of the newly invented vaccuum flask seemed to justify. It became clear that Trouton's rule[42] did not hold for hydrogen. This failure became even more apparent when, ten years later, Kamerlingh Onnes liquefied helium. The extremely low density of the liquid and particularly the density maximum at the λ-point discovered by him in 1911, led him to the vague suggestion that it might be connected with quantum mechanics.[43]

In classical theory the motions of all particles are expected to die out as absolute zero is approached. In quantum mechanics, however, the situation at absolute zero is different: there is a zero-point motion first introduced tentatively in Planck's second hypothesis.[44] He assumed that the energy of an oscillator in its lowest state was $1/2\ h\nu$ (where ν is the frequency of the oscillator). Of course, zero-point energy must affect all properties of condensed phase, and at temperatures for which kT is small in comparison to $h\nu$, this contribution would be considerable. Thus we would not expect zero-point energy to influence appreciably the properties of substances at room temperature, but its effects could become very pronounced in substances with a low boiling point (as they happen to possess low atomic weights and therefore higher frequencies of oscillation).

In 1923, there was not yet any direct evidence for the existence of the zero point energy. In a paper published that year, Bennewitz and Simon (1923) were able to provide this evidence. They pointed out that the deviations of some liquefied gases from Trouton's rule could be explained if the latent heat was reduced by a parameter representing the high zero-point energy of these gases. For solids and liquids composed of atoms of small atomic mass (e.g. hydrogen and helium), in addition to the long-range van der Waals attraction and the short-range repulsion, a third force becomes important: the effective repulsion arising from the zero-point energy due to the atoms being confined in space. Whether or not this zero-point energy is sufficient to hinder the attractive forces to produce solidification depends on the potential energy of molecular attraction. It appears that the conditions for this non-classical behaviour, in which the crystal lattice is disrupted, were fulfilled only in the case of helium, where a large zero-point energy is accompanied by a weak molecular attraction and overcomes it. It was the first time that what was suspected on experimental grounds (that helium remained liquid down to absolute zero) was given further support by this theoretical calculation.

After Keesom's success in 1926 in solidifying helium by applying pressure[45] it became evident that it was not possible to reach the solid phase of helium just by lowering the temperature. It was, thus, deduced that the entropy difference between liquid helium and solid helium tended to zero as the temperature was lowered and this meant that the liquid phase had to go into an ordered state. The nature of this "ordered" liquid state, which Simon called "*liquid degeneracy*",[46] was one of the questions which preoccupied many physicists in the following years.

The fact that helium remained liquid (under its own vapour pressure) down to absolute zero was the first indication that its properties could not be understood in terms of the familiar classical concepts. Were we, in fact, confronted with "pure" quantum effects? Was it the case that we had a macroscopic physical system which was behaving quantum mechanically?

Physicists were confronted with a paradoxical situation: on the one hand, the result (the zero entropy of liquid He II) seemed to be in accordance with the third law of thermodynamics,[47] and, on the other, it could not be reconciled with the dominant notion of what a liquid is, and what constitutes its difference from the solid phase.[48] In order to be able to "reconcile" the two specific contexts (the implications of the third law of the thermodynamics and the implications of the definitions of liquid and solid phases) one would have to examine the very foundations of the "definitions" and "formulations" involved in each context. The observed new phenomenon necessitated the reformulation of classical concepts into forms which took into account the quantum behaviour of matter.

The first attempts to solve the problem were confined to describing liquid helium below T_λ as the extreme case of a "liquid crystal"[49] since the notion of order was so closely related to coordinate space. In such a crystal small crystalline regions of variable size and shape would account for a high degree of order, allowing at the same time the substance *as a whole* to retain its liquid aspect. Keesom called this state "*quasi-crystalline*",[50] and Clusius in an unpublished paper speaks of a "*crystalline*" state.[51]

Three years later, in 1936, the question was taken up in some detail by Fritz London, who in 1933 had moved to Oxford from Berlin with his younger brother Heinz. He had made calculations on the structure of liquid and solid helium to put Simon's ideas on a quantitative basis.[52] Taking into account the large zero point energy of the substance, he compared face-centered, cubic simple cubic and diamond structures and their potential energies as a function of the atomic volume. He found that, under the actual conditions of helium, "the diamond structure would have the lowest energy".[53] He then concluded that if helium II tended to form any of these structures, the diamond lattice would be the one most favoured and posed the following question:

> Is liquid helium below the λ-point to be simply interpreted as going into the ordered state of a diamond lattice?[54]

In his own answer we find the first traces of the new notion of order

> ... *at the best one can speak of a statistical preference for the equilibrium position* ..., and one will gather from our calculations of the energy that *this statistical distribution of the atomic distances* cannot differ very much from that of a diamond lattice ... This conception ... seems not to be so unsatisfactory in view of the very small vicosity of this phase, which is hardly to be understood by assuming that it is simply a *common* crystallized state.[55] (Emphasis added.)

i.e. *the "lattice points" can only give an indication of the position for which the probability of finding an atom would be higher than elsewhere.* Thus, the lattice structure must no longer be considered as rigid, but as preferred configurations of statistical distributions.

Next year, 1937, Fröhlich made the ingenious suggestion that since two diamond lattices shifted by one spacing are equivalent to one body-centred cubic lattice, an order—disorder transition would be possible in which each point of the body-centred lattice is "half occupied" at a temperature above the transition point, but below, the atoms are in *one* of the diamond lattices. Fröhlich suggested that perhaps helium II has the diamond structure and helium I is body-centred. This meant that above the λ-point n helium atoms are distributed at random over the $2n$ positions of the cubic lattice, while below the λ-point they settle down in the n places of one of the diamond "sub"-lattices. This gives about the right entropy change and a discontinuity of the specific heat. F. London criticized this order—disorder theory on the grounds that it was difficult to see why atoms should prefer to settle in *one* of the two available lattices assumed to exist in helium II and concluded that:

> we have a body-centred lattice of $2n$ places for n atoms, *every place being occupied with the probability 1/2 only* — even at the absolute zero.[56]

But what does such a probability mean in quantum mechanics? How does it contribute to the reformulation of the notion of order?

In the ground state a particle is described by a single *wave function which indicates with what probability one can expect to find the particle at a given place.* According to quantum mechanics this is a complete and unique description of the particle. The statistics apply to the wave functions, not to the particle, and *if there is only one single wave function there is no possibility of disorder.* In case of high disorder, there will be a great variety of quantum states. (Above the ground state there is a multitude of excited states which can become occupied at higher temperatures). *The entropy* is defined in such a way that it *is high in the case of many quantum states and zero in the case of only one quantum state.* As F. London emphasized:

> In principle the whole system [of the suggested structure] is supposed to be representable by a *single* quantum state, which if it is the lowest state, will describe the system at the absolute zero. Accordingly, there the system has zero entropy in spite of the fact that it may involve a statistical distribution of N atoms over $2N$ places. *Quantum-mechanical uncertainty holds of course, at absolute zero, but it has nothing to do with thermodynamic disorder and entropy*".[57] (Emphasis added.)

Thus, Fröhlich's argument of an order—disorder transition could still be maintained; only the state of order was one in which n helium atoms settled down in $2n$ half-occupied sites. Here was, then, a situation which indicated how a random distribution can change into an ordered distribution by quantum mechanical principles.[58]

At that time, an X-ray investigation of liquid helium was in progress at Leiden and it yielded for both helium I and II X-ray patterns of the usual diffuse type characteristic of liquids.[59] There appeared, then, to be no real evidence of structural change at the λ-point. It must be borne in mind that many different atomic configurations gave similar X-ray patterns, so that the available experimental data did not rule out the possibility of a change in structure. Also it must be remembered that in a second order transition the structural change would be gradual. In the end, however, all the above considerations of the possible structure of liquid helium brought us no nearer to the explanation of the λ-transformation. In order to explain the transformation one had to look to an *entirely different cooperative phenomenon*, which had nothing to do with crystal structure. If, however, one insisted on having a crystal structure, then the lattice points must mark a series of maxima in the wave-function of every atom; and a particular atom can never be associated with a particular lattice point.

It must be emphasized that the theoretical study of a noncrystalline zero-entropy state was extremely difficult. It would have required the solution of the formidable many-body problem, a rigorous discussion of which had been possible only for the relatively simple case of an ideal gas of non-interacting particles, and one also had to distinguish between two cases, the Fermi—Dirac and the Bose—Einstein statistics. In Fermi statistics, the Pauli exclusion principle is obeyed so that no two identical fermions (e.g. particles having half integer spin) can be in the same quantum state. In Bose—Einstein statistics the Pauli exclusion principle is not obeyed so that the number of identical bosons (e.g. particles having integral spin) can be in the same state.[60]

In contrast to the situation of an ideal gas, molecular interactions were always important for actual systems at low temperatures. Nevertheless, the analysis of the experimental results seemed to indicate that *zero entropy states were characterized by some kind of "ideal" behaviour.* And ideal behaviour patterns (ideal crystals, ideal liquid, ideal conductors or ideal diamagnetic bodies) seem to be characteristic of the pure quantum states in contrast to the high entropy states.

The first six months of 1938 brought a dramatic change in liquid helium theory. The notion of a quasi-crystalline state, "which had never been taken seriously"[61] was given up by F. London himself, who turned to a new model. In this model each helium atom moved nearly freely in the self consistent periodic field formed by the other atoms in a similar way as electrons move in a metal according to Bloch's theory, but with the difference that the helium atoms obey Bose—Einstein (instead of Fermi—Dirac) statistics. As a first step he disregarded the self-consistent field altogether and considered the ideal Bose—Einstein gas.[62] Einstein had in 1925 already discussed a peculiar condensation phenomenon of this gas.[63] Below a certain temperature, which depends on the mass and density of the particles, a finite fraction of them begins to *collect* in the lowest energy state; that is, they assume zero momentum. The remaining particles have a velocity distribution similar to a classical gas, flying about as *individuals*. One might say that the gas breaks up into a mixture of a "condensed" and an "excited" component; both components, however, occupy the total volume of the container as if one dissolved in the other. Thus, there is no condensation in the ordinary sense; that is, there is no separation in space into two phases which can be distinguished by their density. But, "if one likes analogies, one may say that *there is actually a condensation but only in momentum space, and not in ordinary space.*"[64] The condensed particles with their zero momentum are, according to the uncertainty principle, not localized. Each is everywhere. In quantum mechanics they are described by a single wave function which is symmetrical in all particles so that an exchange of two particles leaves the wave function unaltered. Just as was the case with the distribution over two possible lattice positions, we have here a distribution over the whole volume which is described by a single wave function. Therefore the entropy is zero. Thus, *the order assumed at absolute zero is an order in momentum space*. In ordinary space there is no order in the classical sense, but also no disorder; in the quantum mechanical sense, the particles are just no longer localized.

F. London pointed out that such a non-localized structure in condensed helium would, because of the high zero-point energy, be more favourable than a "quasi-crystalline" structure. Indeed, liquid helium II, despite its high degree of "order", instead of being close to a ("liquid" or solid) crystal is, owing to its extremely low density, much closer to a gas than to an ordinary liquid. It was this gas-like nature combined with the high degree of order of helium II which led London to his theory of a condensed Bose—Einstein gas.

For an ideal Bose—Einstein gas the condensation phenomenon actually represents a discontinuity of the derivative of the specific heat (phase transition of third order),[65] and the condensation temperature calculated for the density and mass of liquid helium was 3.09 °K. The actual λ-transition of liquid helium showed a *discontinuity* of the specific heat (phase transition of second order) and took place at 2.19 °K and F. London claimed that:

> Though actually the λ-point of helium resembles rather a phase transition of second order, it seems difficult not to imagine a connection with the Bose—Einstein statistics. . . . On the other hand, it is obvious that [we deal with] a model which is so far away from reality that it simplifies liquid helium to an ideal gas.[66]

Nevertheless, the model could not yield the exact values of the λ-point temperature and its specific heat.

At any rate, the analogy with the Bose—Einstein gas model had for the first time given a clue for the understanding of many of the strange phenomena of liquid helium. An important step in this direction became possible through the two-fluid model proposed by Laszlo Tisza (1938a), an Hungarian-born physicist working at the Collège de France, partly on the basis of F. London's ideas.

F. London's and Tisza's papers, published in this same volume of *Nature*, were actually the first successful trials to impose an order on this "an Alice in Wonderland World".[67]

> The origin of . . . [the two-fluid] concept can be traced to two sources: first, an intuitive evaluation of a situation which can give to such spectacular effects as the superheat-conductivity, superfluidity, the Rollin film, the fountain effect and the apparently hopelessly conflicting viscosity obtained from oscillating disk experiments; second, theoretical ideas, such as the B-E theory in the case of Tisza, and quantum hydrodynamics in the case of Landau.[68]

4.4. The two-fluid model

F. London, who by 1936 had moved to Paris, discussed his work on Bose—Einstein condensation with L. Tisza. Tisza immediately sought ways of utilizing these ideas to devise an explanation to the newly discovered transport phenomena. London was not particularly at ease with Tisza's "initiative" — despite the fact that Tisza never neglected to acknowledge his debt to London. In his note to *Nature*, Tisza proposed what has come to be known as the "two-fluid model" and applied it, with impressive success, to the viscosity paradox and the fountain phenomenon—predicting the mechanocaloric-effect at the same time.

F. London continued to explore the possibilities offered by the idea of Bose—Einstein condensation and published two papers[69] in which he discussed in some detail the theoretical properties of an ideal Bose—Einstein gas with particles equal in mass to helium atoms, and with a molar volume equal to that of liquid helium at the λ-point, but made hardly any reference to the observed phenomena. It did not appear that he thought highly of Tisza's model. And, in fact, eight years later, in a meeting at the Cavendish Laboratory, F. London confessed that:

> I was at that time not very well satisfied with . . . [Tisza's] kind of reasoning. I felt that a matter of this sort should be proved on the basis of the recognised principles of quantum mechanics and quantum statistics and should not be presented as a matter of assumptions or hypotheses. On the other hand, one has to admit that a logical reduction of several apparently independent facts, a macroscopic theory can be a valuable achievement, even if one cannot yet carry everything rigorously back to first principles; one would say this with particular emphasis if the theory in question should prove able to predict the existence of previously unknown phenomena. In fact Tisza's model has been found extremely successful in predicting properties of liquid helium, which at that time were either unknown or merely qualitatively investigated".[70]

Let us consider Tisza's model in some detail.

In this model, helium II is regarded as a mixture of two (completely interpenetrating) components, the normal and the superfluid. These components or fluids are distinguished by different hydrodynamical behaviour, in addition, to the difference in heat content. While the uncondensed normal fluid is supposed to retain the properties of an ordinary liquid (it is identical with helium I), the condensed superfluid fraction of helium II is meant to be incapable of taking part in dissipation processes. At absolute zero, the entire liquid is supposed to be a superfluid consisting of condensed atoms, while at the transition temperature this component vanishes. The hydrodynamical properties of such a mixture are quite complex, but they are flexible enough to explain things which appear puzzling in ordinary hydrodynamics. Hence, an oscillating disc in helium II will experience friction by the normal fluid while a fine capillary will allow superfluid to pass without experiencing friction. Similarly an interpretation could be found for the thermomechanical effect. Since in the model the temperature of a volume of helium II simply means a certain relative concentration of the two fluids, a change in this concentration will be registered either as a cooling or a heating. Absorption of heat has the effect of increasing the concentration of the viscous component and also the osmotic pressure at the expense of the superfluid which is sucked into the cell. The obvious conclusion from this explanation was the prediction of the inverse effect, namely that helium forced through a fine capillary should be richer in superfluid and therefore exhibit a drop in temperature. This effect known as "mechano-caloric effect" was observed in 1939 by Daunt and Mendelssohn (1939b). The anomalously high heat transport in helium II was also consistent with the assumptions of the two-fluid model. The important thing here is that the superfluid and viscous components may have different flow velocities, giving rise to an "internal convection" which is connected with an energy transfer without any mass transfer. This internal convection accounts for the super heat-conductivity.

On the basis of Tisza's hypothesis and general thermodynamics Heinz London gave in 1939 a quantitative theory of thermo-hydrodynamical effects. Assuming the fountain or mechanocaloric effect to be thermodynamically reversible and the flow through the fine capillary to be carried out by a fluid of zero entropy, London arrived at an equation which relates the fountain effect to the entropy of liquid helium.

A few months later, in another short note presented to the Academie des Sciences in Paris, Tisza (1938b) went much further; he recognized that his model implies a very strange feature, namely that in liquid helium II the temperature would obey a wave equation, as does the pressure. Tisza called these waves "temperature waves". When this paper, in which he predicted this peculiar wave propagation ("perhaps the most interesting deduction from the Bose—Einstein theory"[71]), became available it was too late to be tested before war broke out. Work on liquid helium in Western Europe had stopped with the outbreak of World War II, but in the Soviet Union Kapitza was able to detect these waves while performing a number of interesting experiments on critical flow velocities and thermal conduction.[72]

It should be emphasized that Tisza's successful two-fluid model was based on admitting something which was physically impossible: it was possible in his model to distinguish between two fluids, made up of the same species of atoms and which should, therefore, by definition, be indistinguishable. The success of the model often resulted in a tendency to consider the implied mechanism for the superfluid behaviour of helium as a physically real mechanism, and not just an "algorithm" useful in — surely the qualitative — understanding of many of the phenomena associated with superfluid helium. Tisza's two-fluid model had to be modified in such a manner as to be able to recover its qualitative explanations without dividing the atoms into different components, and this was exactly what happened a couple of years later.

In 1941 Kapitza published his papers (1941a, 1941b) containing a large amount of experimental observations on liquid helium II and he proposed ideas which allowed for the possibility of a deeper understanding of the phenomenon of superfluidity and of the mechanism of thermal superconductivity. He put forward a hypothesis claiming that the "abnormal heat conductivity was not due to some exceptional thermal property of helium II but to heat transfered by convection currents whose presence can be anticipated owing to the exceptionally high fluidity of helium II".[73] However, a closer quantitative analysis of the experimental evidence revealed new difficulties.

Calculations showed that to explain the thermal conductivity values observed by Keesom, and Keesom (1936), the convection velocity must be assumed to be about 50 m/s. This is a considerable velocity indeed and Kapitza decided to measure it more accurately. For this purpose he set up a series of experiments in which he observed a heat transfer at least 20 times greater than that measured by Keesom. Consequently, the convection velocity necessary for explaining such thermal conductivity would have to be of the order of 1000 m/s rather than the expected 50 m/s. Velocities of such magnitude do not exist in convection currents,[74] and this result led to further difficulties with the then accepted two mechanisms for heat transfer. One of these mechanisms assumed convection and the other was the conventional mechanism of thermal conductivity involving the transmission of thermal motion from one atom to the next. From Kapitza's measurements of convection velocity it became clear that the first mechanism had to be ruled out as a possible mechanism for the specific kind of heat transfer. In view of Kapitza's results, the second mechanism had to be ruled out too. Indeed, if a layer of atoms is suddenly heated, the atoms involved will oscillate and these oscillations will be transmitted from one layer to the next in the form of a heat wave. It was found that the velocity of such a heat cannot exceed that of elastic oscillations in the body (i.e. the velocity of sound). The velocity of sound in helium II was found to be 230 m/s, while the velocity obtained in thermal measurements was several times greater.

Here is Kapitza's own description of his attempts to solve these contradictions in the absence of any guiding theory:

> We spent about a year in attempts to solve these contradictions. How could we proceed with our search for the true mechanism of this heat transfer in the absence of any guiding thought? Indeed our findings basically contradicted all the established theoretical concepts. We had to grope along, testing various physical factors which might affect the thermal conductivity. Thus we examined the heat transfer of helium II as a function of pressure, gravity, time, etc. Only negative results were obtained; the thermal conductivity remaining unchanged at its high level. Finally an accidental observation gave our work an impetus in a totally new direction.[75]

He found that pressure pulsations transmitted quite accidentally from the helium pipeline of the laboratory into the helium in the capillary caused substantial changes in the thermal conductivity. In order to study this phenomenon as fully as possible, Kapitza set up new experiments, and thus he was able to establish the mechanism of the movement of liquid helium in the capillary as a result of a heat current.

Kapitza proposed an explanation in which he did not make use of the two fluid model. Instead he suggested the possibility of two spatially separated mass currents, flowing into the bulb of the surface layer on the inner perimeter of the tube and outflowing through the centre of the tube. In order to explain the great thermal conductivity of helium II on the basis of this pattern of movement, he suggested that there is a difference between the heat function of helium in this film and in the free state, and thus the different in heat content between the two mass currents was accounted for by the van der Waals forces of the capillary wall on the surface layer of liquid. This hypothesis proved fruitful; it led to the prediction that the thermal conductivity of helium would be strictly normal in the absence of surface phenomena. From his subsequent experiments Kapitza (1941b) concluded that the entropy of liquid helium flowing through a narrow channel was zero, stating that

> the possibility of such a state with zero entropy for helium II has already been suggested by Tisza and London but up to now it had no rigid theoretical basis.[76]

As the more likely theory to provide such a basis Kapitza pointed to the theory of liquid helium by Landau which was published simultaneously with his own paper.[77]

4.5. Towards a microscopic theory

Although by the mid nineteen forties there was no completely satisfactory theory of the liquid state, the theories of the gaseous and solid states were well established, and it was therefore tempting to treat a liquid either as a very imperfect gas in which the intermolecular forces have become important or as a broken down solid in which the binding forces are too weak to localize the atoms near the lattice points.[78]

Lev Landau rejected the idea that the type of statistics obeyed by the helium atoms had anything to do with the superfluid properties of helium II, and approached the problem from a point of view similar to that used to develop a theory of the solid state. In the theory of Bose—Einstein condensation, emphasis was placed on the wave-function of the individual atoms. In the theory of solids, it is not possible to give the wave function for an *individual* atom, but one discusses instead the normal modes of motion of the solid as a *whole*. In Landau's theory the problem of accounting for the interaction forces between the helium atoms was avoided by treating the liquid as a quasi-continuum. He attempted to construct a quantum theory of liquids by the direct quantization of the hydrodynamical variables such as the density, the current and the velocity, without explicit reference to the interatomic forces.[79] He considered the quantized states of motion of the whole liquid instead of the states of the single atoms, and started by considering the state of the fluid at absolute zero, which is its ground state.[80] Thus excitation of vorticity would represent departure from the zero temperature states. Departure from the ground state could also arise from the excitation of one or more units of sound-wave energy, or "phonons", which give rise to a Debye-type specific heat and could not account for the specific heat above 1 °K. In this way, Landau constructed the energy spectrum of liquid from two types of excitations; to the phonons of the solid body he added a spectrum of "rotons" by which term he defined the elementary excitations of the vortex spectrum.[81] States near the ground state, therefore, were characterised by the numbers and energies of the phonons and rotons superimposed on the ground state.

Therefore, according to Landau, one might picture the helium as a background fluid in which excitations move, and thus there exists just one fluid: liquid helium. At the same time, one can say that the ground state and the excitations, respectively, play the role of superfluid and normal fluid.[82] The excitations are "normal" because they may be scattered and reflected, and hence show viscosity. The fluid associated with the ground state is superfluid because it could not absorb a phonon from the walls of the tube or a roton unless it was flowing with a velocity greater than the velocity of sound or a "critical velocity", respectively. Below the lesser of these two velocities the flowing helium would not interact with the walls and, hence, would be superfluid (unless, as Landau pointed out, some other mechanism, as yet undetermined, limited the flow).

Landau's 1941 paper marked the beginning of a series of disputes between him and London and Tisza. Landau himself claimed that Tisza's point of view "cannot be considered as satisfactory. Apart from the fact that liquid helium has nothing to do with an ideal gas, atoms in the normal state would not behave as "superfluid" . . . the explanation advanced by Tisza not only has no foundation in his suggestions but is in direct contradiction with them";[83] Ginzburg stated that "the condensation of a Bose gas has nothing to do with the properties of helium. The latter have been explained by Landau's theory only",[84] and Peshkov rejected the London—Tisza theory as "very artificial and unconvincing".[85] F. London and Tisza responded by pointing out that "the B—E gas, at any rate, has provided an idealized model for nearly all the unique properties of liquid helium, which model has furthermore led to predictions of new effects,"[86] and that "one has to give the preference to the B—E theory over the quantum hydrodynamic approach".[87]

London's absolute insistence on the significance of statistics for explaining superfluidity is found throughout his articles and it is strongly expressed in his correspondence mainly with Tisza. Tisza, much more than London, seems to have realized the threat that the "second sound" measurements below 1 °K, posed for the two-fluid model.

An experimental corroboration of the essential validity of the two-fluid model of helium II, as stated by Landau, was Andronikashvili's (1946) measurements of the superfluid fraction. The experiment consisted in measuring the moment of inertia of a pile of closely-spaced aluminium plates suspended in a bath of liquid helium II. Since the superfluid component is capable only of irrotational flow it has no effect on the rotation of the disks. The normal component will in part be carried around with the movement of the discs when they are set in oscillatory motion. Indeed the rotation of the liquid proved much less than would normally be expected, indicating that only the normal component of the helium rotated, while the superfluid remained at rest. By measuring the variation in the period of oscillation with change in temperature, it is possible to calculate the variation of the total moment of inertia of the oscillating system and hence obtain the relative density of the normal component present at any given temperature. The result showed that the superfluid fraction varies from 0 at T_λ to about 70% at 1.76°K which means that above the lambda point the liquid is entirely normal and that the "amount" of the superfluid increases as the temperature drops below T_λ.

The formalism of Landau's theory led to two different equations for the propagation of sound, and hence to two velocities of sound. One of them was related in the usual way to compressibility while the other one depended strongly on temperature. Such formalism was the reason for giving the name of "second sound" to Tisza's thermal or temperature waves.[88] Indeed the first and unsuccessful attempt to generate and detect second sound waves was made with acoustic apparatus by Shalnikov and Sokolov.[89] Their failure, as Brush notes in his excellent treatise on the development of the ideas in statistical physics and the atomic structure of matter

> meant that Landau's theory was "born refuted": it could not account for any properties of helium II that were not already explained by the London—Tisza theory; it offered no explanation of the λ-transition; its prediction for specific heat and critical velocity were in disagreement with experiments; and the expected property of "second sound" propagation had not been observed.[90]

The failure to observe second sound acoustically was explained in 1944 by another of Landau's colleagues, E. M. Lifshitz who made a more detailed theoretical analysis of second sound waves and showed that "Radiation from a surface whose temperature fluctuates periodically in time is even more favourable for the "second sound"."[91] Using such a radiator, Peshkov was able to demonstrate in the same year the existence of standing thermal waves for the first time.[92]

Both theories of Landau and Tisza predicted a zero value for the "second sound" velocity at the λ-point, and both predicted the same shape of curve down to 1°K. Below that temperature the theories diverge, Tisza's predicting a zero velocity at 0°K and Landau's predicting that velocity should rise again steeply after a shallow minimum to a constant value at 0°K. In 1946, having further perfected his techniques, Peshkov published more detailed results[93] and concluded that

The temperature dependence of the velocity of the second sound fully confirms the phenomenological part of the theory of superfluidity and in particular the hydrodynamics of helium II, which predicts the possibility of two simultaneous motions the normal and superfluid ones. The microscopic theory agrees with the experiments less satisfactorily.[94]

Landau, in 1947, modified the roton energy spectrum and proposed to combine both phonon and roton excitations in a single energy-momentum curve. This modification was introduced in order to bring his evaluation for the velocity of second sound into better agreement with Peshkov's experimental results.[95] With such a spectrum, Landau said,

> it is, of course, impossible to speak strictly of rotons and phonons as of quantitatively different types of elementary excitations. It would be more correct to speak simply of the long wave . . . and short wave . . . excitations,[96]

or, equivalently, about the phonon and roton part of the dispersion curve but, he stressed that

> all the conclusions concerning the superfluidity and the entire macroscopical hydrodynamics of helium II, developed in [1941 paper] . . . maintain their validity also with the spectrum proposed here[97]

Tisza (1947) criticized Landau's modification of his theory, but at that time, Bogoliubov found that the energy spectrum could be expressed in terms of "quasi-particles", representing the elementary excitations. Bogoliubov concluded that "no division of quasi-particles into two different types, phonons and rotons, can even be spoken of"[98] thus favouring the recent modification of Landau's theory.

The observations to temperatures below 1 °K (where Landau predicted roton contributions to become effective) was in favour of the Landau treatment and against the approach to regard liquid helium as an ideal Bose-Einstein gas. Despite Tisza's unsubstantiated claim that "there seems to be no difficulty in incorporating . . . the experimental results below 1 °K into the B-E theory",[99] there was no convincing way of modifying the two-fluid model to account for the observed measurements. The discussion for a short time afterward, concentrated on questions pertaining to the foundations of the two approaches and the role of symmetries. The new consensus was reflected by the attitude of one of the pioneers of the field, J. F. Allen

> liquid helium *is* a liquid and that we must be wary of a description in which certain atoms are labelled "superfluid" atoms while others are "normal" atoms in excited states. Whatever may be the true natures of the superfluid and

> normal fluid, their method of excitation must as we see it at present, be described in terms of a spectrum such as Landau describes.[100]

The question, however, of whether the Bose—Einstein condensation was essential for a theory of superfluidity, did not seem to have been settled.

At that time "a new possibility has opened for deciding, by experiment, the relevancy of the Bose—Einstein way of counting, without entering into the obscurities of complicated and approximate calculations".[101] Ordinary helium has an extremely rare isotope (He^3), which obeys Fermi—Dirac statistics and it had been assumed that if the Bose—Einstein statistics were essential to an understanding of the superfluidity of the isotope He^4, then He^3 should be not superfluid. Since He^3 is present in ordinary helium only as one part in ten million, experiments were very difficult. Preliminary experiments did, in fact, indicate that He^3 was not superfluid but "one might hold defining judgment in suspense until the experiment is extended to a still lower temperature."[102] And such was, in fact, the case at about 2.5×10^{-2}°K when liquid He^3 was also found to be superfluid.

Tisza and Landau provided the most appropriate theoretical tools to handle the problem of the extraordinary behaviour of helium II. By developing the concepts of "superfluid" or"background fluid" and of "normal fluid" or "elementary excitations" they circumvented the difficulties associated with a description in terms of individual atoms which do not move independently. These intermediate concepts[103] made it possible to think in simple terms about the liquid helium system, they provided insights into the behaviour expected of the system and, moreover, allowed qualitative predictions to be made about it (thermal waves or second sound).

A possible connection between the Landau theory of superfluidity and the Bose—Einstein condensation was demonstrated by Bogoliubov. His work was a major step towards unifying the London—Tisza and Landau approaches. He attempted to construct a consistent molecular theory explaining the phenomenon of superfluidity without assumptions concerning the structure of the energy spectrum. According to him

> The most natural starting point for such a theory seems to be the scheme of a nonperfect Bose—Einstein gas with a weak interaction between its particles.[104]

By using the method of second quantization, together with an approximation procedure, Bogoliubov showed that in the case of weak interactions between molecules the low excited states of the gas can be described as a perfect Bose—Einstein gas of "quasi-particles", representing the elementary excitations, whose "existence and properties . . . follow directly from the basic equation describing the Bose—Einstein condensation of non-perfect gases . . ."[105] He also found that even at zero temperature not all molecules have zero momentum, and because of this, one could no longer simply identify the superfluid fraction of particles in the zero-momentum state, as was done in the original London-Tisza theory. "Thus in the unification of the two theories each had to give up some of its original features".[106]

Although the interaction between helium molecules is comparatively weak, liquid helium cannot be considered as a "nearly perfect" Bose—Einstein gas. So the justification of the Landau spectrum on the basis of a microscopic theory remained still open. As Bogoliubov himself wrote in 1947,

> a rigorous theoretical computation of the properties of a real liquid is hopelessly beyond the reach of a pure molecular theory based on usual "microscopic" equations of quantum mechanics. All we can require from a molecular theory of superfluidity, at least at the first stage of investigation, is to be able to account for the qualitative picture of this phenomenon being based on a certain simplified scheme.[107]

In 1952, Temperley, in view of previous experimental results, concluded that the Landau approach was the proper one for the lowest temperatures, but that it loses its validity above, say, 0.6 °K while the London—Tisza model is valid above, say, 1.5 °K Between these temperatures there is a transition region in which neither a solid nor an almost perfect gas furnishes a good approximate model. In attempting to reconcile the two models (London—Tisza and Landau), Temperley outlined a theory in which large clusters in coordinate space are in statistical equilibrium with small clusters in momentum space which play the role of the superfluid component. The λ transition expressed the appearance of the clusters in momentum space. The thermodynamics resulting from the assumptions of the occurence of molecular association below T_λ had been discussed by Rice, but measurements of the electric polarizability, of the refractivity and of the light scattering failed to reveal any significant difference in these quantities either in helium I or helium II. Rice, moreover, suggested that there was some reason to suppose that the two fluids of helium are separated not only in momentum space as in F. London's theory but also in ordinary space.[108]

All these theoretical considerations[109] justified in some way the two-fluid model, and elucidated some properties of helium II, but did not provide a successful microscopic theory of liquid helium. F. London described the situation, in 1951, as follows:

> Five years ago I was able to report to the Low Temperature Conference in Cambridge that the theory of Liquid Helium seemed to be in a very satisfactory state . . . the two-fluid model . . . could explain or predict most of the striking properties of He II . . . . Today the position is not quite so simple. No satisfactory molecular theory of liquid helium has so far been produced and

> there are evidently limits to the validity of the macroscopic two-fluid model, although the general picture seems to be nearly correct.[110]

It was Feynman a couple of years later who showed that one can justify the two-fluid model quantum mechanics applied at the atomic level and remedied, to a large extent, its major failures.

He embarked upon a program with the purpose of answering the following questions: Why does the liquid make a transition between two forms, HeI and HeII? Why are there no states of very low energy, other than the phonons, which can be excited in HeII at very low temperatures? What is the nature of the excitations which constitute the "normal fluid component" from 1 to 2.2 °K?

Feynman published a series of papers in which his views on the matter had been successively developed. In his excellent summary of his final position we read:

> There is . . . little doubt that in liquid helium there are such excitations, with the energy spectrum that Landau suggests, and that this picture supplies the complete interpretation of the two fluid model for helium II . . . The next question that concerns us is to try to see from first principles why the excitations of the helium fluid have these characteristics . . . Landau has, in fact, tried to obtain some justification for the spectrum from a study of quantum hydrodynamics. This is not a completely detailed atomic approach. However *it is possible from first principles to see why there are no other excitations but those supposed by Landau and why the energy spectrum of these excitation has, qualitatively, the form which he supposed.*[111] (Emphasis added.)

In his first paper Feynman attempted to show that the interatomic potential did not change the existence of an Einstein—Bose condensation in He^4.[112] The idea was that, since one is dealing with a loose liquid structure, the motion of an atom is not opposed by potential barriers because the other atoms can move out of the way to make room for it. This readjustment of the positions of the other atoms merely increases the effective mass of the moving helium atom, and brings us to something very like a gas of hard spheres (having effective masses $m' > m$). The mathematical situation turns out to be very similar to that which gives rise to the Bose—Einstein condensation of an ideal gas. A third order transition occurs and the rise in specific heat just above the λ-point is qualitatively explained. However, the theory predicts a specific heat varying as $T^{3/2}$ near 0 °K, badly at variance with the observed T^3 variation.

The difficulty of the low temperature specific heat led Feynman in his subsequent paper to a consideration of the low energy excitations which exist in the liquid near 0 °K. He showed that the excited state is similar to the ground state with a small perturbation (the phonon). A long wavelength phonon involves the motion of many atoms, each of which moves only a short distance, and the neglect of such motions were responsible for the absence of the T^3 specific heat term in his first paper. Moreover, he eliminated the $T^{3/2}$ specific heat term by showing that the phonons are the only low lying states.[113] In a subsequent paper[114] the arguments were extended to give quantitative results about the spectrum of excitations.

The Landau and Feynman curves have similar shapes, but the numerical agreement is poor and Feynman, with Cohen, attempted to find a better wave function. The new wave function permitted an improved agreement with experiment, and suggested that the roton is the quantum mechanical analogue of a microscopic vortex ring.[115]

As we have already mentioned, one of the major failures of Landau's two-fluid model, as compared with atomistic theories based directly on Bose—Einstein statistics, has been its treatment of rotational motion. The disagreement between the theory and the results of experiments on both the free surface of rotating helium and on the critical velocity in helium was essentially resolved by the hypothesis about the existence of vortex sheets and especially about the vortex lines and quantization of the superfluid circulation in helium proposed by Onsager[116] and developed by Feynman.

> In ordinary fluids flowing rapidly and with very low viscosity the phenomena of turbulence sets in.
>
> ... The resistance to flow somewhat above critical velocity must be the analogue in superfluid helium of turbulence, and a close analogue at that.
>
> There are some ways, however, in which the two cases differ. In a classical fluid ... viscosity is the mechanism which determines whether vorticity will be amplified or not, and therefore whether turbulence is produced. If the viscosity goes to zero as a limit (and no other physical phenomena are added) a classical ideal liquid would exhibit turbulence at any velocity, no matter how small.
>
> Superfluid helium is an ideal fluid of zero viscosity. It does not exhibit turbulence at low velocity because of another, quantum mechanical, effect. The vorticity is quantized and cannot begin at as low an amount as desired. One must supply energy enough to get the first one or two vortex lines started before the amplification process of turbulence can take over".[117]

Following this assumption Feynman estimated the magnitude of the critical velocity to be 100 cm/s. This result was in good agreement with the experimental value (70 cm/s).[118]

Thus, the failures of the Landau model were to a large extent remedied by the work of Feynman, who placed the whole problem of the excitations in liquid helium on a quantum mechanical basis.

PART III

The therefore

One wonders . . . how Herodotus could believe in the oracle of Delphi, in his time, as he was an intelligent man. What really happens is that each of the predictions of the oracle are in a vague language and they become particularly clear when the event occurs afterwards, so you see how it works. The high priests of Babylon used to predict things by looking at the liver of a sheep. Any why? Because in the complexity of the arrangement of the veins, interpreted correctly, they could tell what the future would be. It is that complexity and the possibility of re-interpreting later the arrangement of the veins, that permits the power of the priests to be maintained.

R. Feynman

CHAPTER 5

(Re-)reading the developments

5.1. Searching for the right problem

The abrupt drop of the electrical resistance of mercury to almost zero, as the temperature was lowered below 4.2 °K, was indeed an unexpected phenomenon, but one which did not, however, establish a paradoxical situation.[1] The theories and models of electrical conductivity during the time superconductivity was discovered did not all have a unique prediction for the behaviour of the electrical resistance at very low temperatures, and the theoretical framework of the period did allow electrical resistances to become negligibly small.

What eventually established a paradoxical situation was the basic assumption behind the initial efforts to find an explanation for the observed new phenomenon. A superconductor was considered to be a *perfect* conductor and, even though Kamerlingh Onnes from the beginning suspected that quantum effects caused such a peculiar behaviour, the phenomenon was not regarded as a "pure quantum phenomenon", but rather as the manifestation of quantum effects in an otherwise classically perfect conductor.

Kamerlingh Onnes' attempts to find an explanation continued during the first couple of years after the discovery of superconductivity, and his efforts displayed his characteristic methodology. He initially accepted Lorentz's electron theory supplemented by the hypothesis that the resistance was caused by the Planck oscillators and that the electrons move freely through the atoms as long as they do not collide with the oscillators and are reflected as perfectly elastic bodies at the surface of the conductor. If one adds to all this that the distance travelled by electrons between two collisions was assumed to be about the size of the conductor and that their drift velocity could not be neglected when compared to the velocity of the "heat movement", then it may have seemed possible to explain the deviations from Ohm's law for the very low temperatures. Despite such theorising, Onnes did not jump to any premature conclusions, stating very clearly his overall attitude on such matters:

As long as the contrary is not experimentally proved, we shall, however, adhere to this law, because we have first to try to refer the phenomena as much as possible to already known ones and so far under appropriate suppositions from the domain of known phenomena the results obtained did not seem incompatible with Ohm's law.[2]

The experiments of 1913 not only established beyond doubt the new phenomenon of superconductivity, they also provided extremely important clues on the method followed by Onnes in his researches — a method never explicitly mentioned in its details but followed consistently throughout his investigations. This method became even more pronounced during periods of "crucial experimentation" where experimental measurements were followed by theoretical speculations, and theoretical assumptions were tested by further experiments. More analytically, one becomes convinced of the following: after a phenomenon was established, experiments were planned to measure specific aspects of the phenomenon and to test various preliminary assumptions adopted for its explanation. These assumptions always took into consideration certain quantum mechanical concepts, but they were at the same time "highly" classical in the overall picture they provided for the underlying mechanism. Moreover, these assumptions attempted, through analogy, to "extrapolate" the validity of the laws used for normal states so as to include extreme cases such as the superconducting states. Phenomenological formulae were constructed and tested for different materials. When it came to testing various assumptions about possible models, most of Onnes' papers concluded not by any strong adherence to a specific assumption — even in cases where the experimental data seemed to support the assumptions. These papers usually concluded with ideas containing new and untested assumptions, and proposals for the kinds of new experiments that would test these new assumptions.

Now, of course, we know that no successful theory of superconductivity could have been proposed before the full development of quantum mechanics by 1926 and the construction of a quantum mechanical theory of electrical conduction in 1928. But superconductivity defied a solution until 1933—34 and the first successful explanation followed right after the discovery of the diamagnetic character of superconductivity — a situation supportive of our claim that what hindered the development of a successful explanation was that the scientific community was working on the wrong problem. It is characteristic that as late as 1931, emphasis was still being placed on that aspect of superconductivity which proved to be such a hindrance to the understanding of the underlying mechanisms of this phenomenon. "We therefore regard the vanishing of the resistance within a few hundrendths of a degree as the most characteristic phenomenon of superconductivity in pure metals."[3]

The problem was formulated within a framework whose limits of validity were determined by the expected behaviour of metals at low temperatures through the extrapolation of their electrical properties at normal temperatures, while leaving their magnetic properties invariant. The invariance of the magnetic properties as the temperature was lowered and as the superconducting state was reached, which implied that the phenomenon was irreversible, was adhered to for no other reason than a strong belief in the Maxwell equations and the mistaken interpretation of Kamerlingh Onnes' and Tuyn's experiment. This outlook did not allow dealing with the phenomenon thermodynamically since such an approach could only be applied to reversible processes.

During this first period, the formulation of the problem did not establish a paradoxical situation within the theoretical and conceptual framework used for problems regarding solids. One was confronted with an "electrical problem", and an explanation was sought for the *sudden* change in electrical conductivity. Thus, the outstanding difficulty with the new phenomenon was to understand its unexpected rate of change and its abrupt drop in electrical resistance. The referential property of superconductors was believed to be their constitutive property as well, a belief that subsequent developments rendered indefensible.

It is interesting to note that right after the development of Bloch's theory for electrical conduction in 1928, a very peculiar situation arose which, in retrospect at least, could have been considered an indication as to the inadequacy of the view that superconductivity is a perfect conductor. Bloch proved a theorem stating that the lowest energy state of the electrons in a conductor (in the absence of an external field) was the state with zero current. Therefore, there was the embarrassing generalization that "every theory of superconductivity is wrong": what was considered to be a superconductor could not be compatible with its observed behavior. Nobody appears to have realized the significance of Bloch's theorem: It was hinting at a different formulation of the physical problem. The community of physicists, like it did in so many other instances, arranged a symbiotic relationship with this "uneasy" result.

The first thermodynamical calculations gave results which were consistent with the experimentally measured values, but the whole approach could not be justified within the framework used to formulate the problem of superconductivity. The problem of infinite conductivity had to be solved while a superconductor was assumed to have the magnetic properties of metals as predicted for perfect conductors from Maxwell's equations — something which did not allow the use of thermodynamic arguments. Keesom, Rutgers and Gorter's reaction was to propose a solution along thermodynamical lines, overlooking, for the time being, the incompatibility with the implications of the framework used to formulate the problem of superconductivity.

The Meissner effect established in an unambiguous manner that the state of a conductor, which at some temperature becomes superconductive, so far as its magnetic properties were concerned, was independent of its past history. Thus, the use of thermodynamic arguments became "legitimized" since one was now dealing with a reversible phenomenon. Superconductivity, it turned out, had a constitutive property distinctly different from its referential property.

Schematically, the problem of superconductivity was transformed from a predominantly electrical problem into a magnetic one. "It seems to me", Bardeen said,

> that most of those who thought long and hard about superconductivity prior to the discovery of the Meissner effect in 1933 never got over an inner feeling that the really fundamental property of a superconductor is infinite conductivity or persistent currents and this coloured the way they thought about the subject in future years. While an adequate theory must explain both aspects, the diamagnetic approach has been the most fundamental in indicating the nature of the superconducting state.[4]

This shift from the electrical to the magnetic problem occurred in a peculiarly dual way. First, starting with the De Haas—Voogd—Keesom results, Ehrenfest's suggestion, and the Rutgers — Gorter approaches, there seemed to have been a heuristic suggestive of a theoretical treatment which would not be confined by the demands of the theoretical framework used to define the terms of the (electrical) problem.

Secondly, the Meissner effect not only established the diamagnetic properties of the superconductors, it also established a paradoxical situation within the then dominant theoretical framework since it was that framework which stipulated that lowering the temperature cannot cause the creation of a diamagnetic state. The outlook suddenly reversed itself from what it had been a few years earlier. With the creation of a paradoxical situation and the subsequent formulation of the "right" problem, a satisfactory explanation was to follow in a short while.

The experimental demonstration that the Maxwell equations did not have the "expected" behaviour at low temperatures had a further methodological outcome: the Maxwell theory appeared to be limited, and the classical framework, even after its "modifications" by the Special Relativity Theory, which to many peoples' minds vindicated electromagnetic theory, was found to require a reexamination in terms of its "natural interpretations".

The historical and methodological significance of the discovery of the Meissner effect was not confined to falsifying a very specific "prediction" of the Maxwell equations or to questioning their extrapolation in order to deduce the magnetic properties of perfect conductors or even to creating a paradoxical situation resulting in the formulation of the (right) physical problem which later was to be explained phenomenologically. The discovery of the diamagnetic character of superconductors was another dramatic instance obliging the community to rethink the criteria by which some of the results of such extrapolations come to be accepted, and why it is that there is a class of results which one doesn't bother to test experimentally; yet we cling to them with such tremendous faith, while others are subjected to the most thorough tests. It would be wrong to reduce such behavior exclusively to factors within the sociology of the scientific community (as it would also be wrong to overlook those factors while attempting to get an insight irrespective of this behaviour). The point we want to make here has a more modest aim. It is about the metaphysical beliefs of the community and its ideological assumptoins: it is usually these beliefs and assumptions, invariably formed and strengthened by the success of a major theory, which become so decisive in the way the results of "extrapolations" come to be accepted as "quite natural" without any need for experimental testing.

The discovery of the Meissner effect did show that it was the faith in the Maxwell equations, together with a specific attitude of what a perfect conductor is, which determined the expectations about the magnetic character of superconductors. The discovery of the Meissner effect undermined a common attitude of the time which held that the extrapolation of classical cases to certain extremes would yield the classical case but in its "idealized form" and that the observed deviations would be due to quantum perturbations. It was the developments in low temperature physics, and especially those in superconductivity and superfluidity which helped change this attitude. And it is this change of attitude that led to a satisfactory phenomenological theory of superconductivity and to the seemingly simplistic model of superfluidity.

The "problem", however, with the two-fluid model was that it was too readily visualizable: One had only to assume the existence of two different and interpenetrating fluids of the same substance with a — quantum mechanical no doubt — mechanism by which one was transformed into the other and where all (classical) hydrodynamical effects were absent. Understanding this "unphysical" situation motivated many researchers to get involved in a particularly productive process of "concepts out of context(s)", thus (re)interpreting many of the (classical) hydrodynamical concepts. One wonders, though, what would have happened if there had been from the beginning, a realization that the two-fluid model does not involve a physical separation of the two fluids, but rather a "separation" that makes the mathematical treatment of the problem more convenient.

Dingle presents a particularly insightful critique of the accepted terminology:

> The terms "two-fluid theory" or "two-fluid model" are, in my opinion, unfortunate . . . The significant model is not so much that of two fluids, but that corresponding to the choice of the excitations, and in the supposition that their mutual interactions do not appreciably influence the results, an assumption also made in the usual calculations of thermodynamical quantities. The rest of the theory is concerned with the correlations between the fluctuations in the density in the first sound, and in the entropy in second sound.
>
> It seems to me that the opinion sometimes expressed that the theory

assumes that there is no interchange of momentum between the two fluids is to some extent due to a misinterpretation. Deviations from the ground state configuration are described in terms of excitations, and most interactions within the assembly can therefore be expressed in terms of the properties of those excitations, such as their mean free path and changes in their number. On this picture, the superfluid just represents the background through which the excitations forming the normal fluid move.[5]

It was this "physical impossibility" together with the "mathematical convenience" of the two-fluid model which prompted some of the developments in (re)determining the conceptual background for solving the problem of superconductivity as well.

The existence of superfluids and in particular the strange transfer mechanisms they exhibit are direct indications that *they represent macroscopic systems* for which the classical theory is incompetent and that presumably quantum mechanics is relevant to their constitution as a whole. In recent years more and more convincing evidence has been brought forward to support the idea that these low temperature transfer mechanisms are *pure quantum mechanisms.* We mean that they are, although of a macroscopic size, nevertheless withdrawn from the disorder of thermal agitation in essentially the same manner as is the electronic motion within atoms and molecules, in short, that they *are quantum mechanics of macroscopic scale*"[6]. (Emphasis added.)

From 1933 to 1950, our understanding of superconductivity was based on thermodynamic and phenomenological arguments. The equations derived from thermodynamics were the only reliable theoretical relations known for superconductors.

F. London was the first to draw attention to the fundamental implications of the Meissner—Ochsenfeld experiment and to develop a set of descriptive, phenomenological equations for superconductors suggested by this experiment. The immediate reaction to the Meissner effect was to try to fit it into Maxwell's electrodynamics, but, with the permeability changing to zero, the equation became indeterminate. It was F. London who came to the conclusion that superconductivity demanded an entirely new relation in which the current is connected not with the electric but with the magnetic field. Together with his brother Heinz London, he proposed that the connection between magnetic field **H** and current density j for the pure superconducting case may be given by the equation (6). (see Chapter 3)

London's equation (6) can be obtained by time integration from (3), if it is assumed that the constant of integration in zero. The assumption $\mathbf{H}_0 = 0$ is, of course, arbitrary (but the exclusion of other possible values of $\mathbf{H}_0$ is in accordance with London's idea of the superconductor as a macroscopic "quantum mechanism"). For this reason London devoted himself to the task of assigning to his equation a "true" physical meaning. The primary importance of such an interpretation is that it frequently leads to a suggestive research program and growth, and this is exactly what happened in this case. The discussion of the implications and the background of the equation stressed the diamagnetic aspect of superconductivity, while the complete disappearance of the electrical resistance appeared, at most, as a secondary quality.

Equations (2) and (6) provide an adequate description of a macroscopic superconductor. They describe the zero resistance and the Meissner effect respectively, and it is obviously important to decide whether it is at all possible to say that one of these effects is a consequence of the other or whether they must be regarded as entirely separate phenomena requiring separate explanations in a fundamental theory.

Equation (6) says more than (3), so far as it includes the Meissner effect. Proceeding from (6) to (3) by differentiating with respect to time we lose this content and we cannot deduce (2). We obtain from (3) the weaker statement:

$$\operatorname{curl}\left(4n\lambda^2/c^2\, j - E\right) = 0$$

which shows only that

$$4n\lambda^2/c^2\, j - E = \operatorname{grad}\mu, \quad \text{or} \quad 4n\lambda^2/c^2\,(j - \operatorname{grad} c^2\mu/4n\lambda^2) = E, \qquad (7)$$

where μ is a scalar. On the other hand (2) leads not to (6) but only to its time derivative (3). Thus, the propositions (2) and (6) "possess, so to speak, the same degree of generality".[7]

Now the question arises whether μ in (7) is merely an arbitrarily chosen integration constant or whether it represents a quantity with a specific physical significance. London, comparing (6) and (7), put them together in the form of an equation for an antisymmetrical tensor. Then the quantity $c^2\mu/4n\lambda^2$ has to be regarded as the time-like supplement of the current density, i.e., the density of charge. This interpretation gave to equation (7) the quality of an independent physical statement.

The logical relation between the three propositions (2), (6), and (7) may now be represented by the scheme:

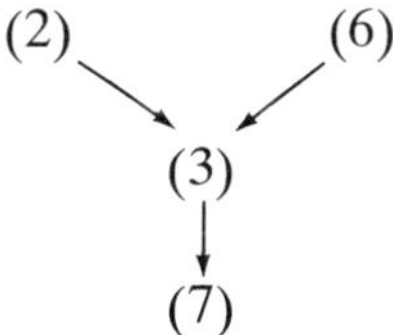

From the above analysis, it is obvious that (2) cannot be deduced from — and does not lead to — (6); they both lead only to equation (3) and to the weaker statement (7). Inasmuch as (2) says more than (3), it expresses a prejudice which is not tested by experience; "it gives nature more freedom that it wants".[8]

In fact, for a simply-connected body (2) can be deduced from (6) though the reverse is not possible. But, it is for a multiply-connected body such as a ring that it is important to prove (2), and this turns out to be impossible without further hypotheses.

It is not, then, unreasonable to take (6) to be "more fundamental" than (2), and this is probably an indication that a supercurrent is to be regarded as a kind of diamagnetic current. F. London, trying to throw some light on the relation between the behavior of a ring (i.e., zero resistance) and the Meissner effect, showned that (6) can be expressed in a rather more significant way, thereby providing a clue as to what is required of a fundamental theory of superconductivity. He suggested that the entire superconductor behaves as a "single big diamagnetic atom". He then went on to argue that if the ground state eigenfunction is 'rigid' and thus not modified very much by an applied magnetic field, the current density will be proportional to the vector potential and thus give the equation which describes the Meissner effect.

The situation can be better described by pointing out that London's original *ad hoc* formula — which fitted and corrected Meissner's data — could be explained progressively within the new theoretical programme, while his "*ad hoc*" formula could not be explained within the phenomenological programme except at the price of a degenerating problemshift. The progressiveness of the new development was abundantly clear: it predicted correctly the value of the fields (not only in the interior of a macroscopic specimen but also in the region close to the surface, and in specimens of dimensions comparable to λ) and a series of other novel facts. But before the invention of the new — but *ad hoc* — auxiliary hypotheses in the old programme, before the unfolding of the new programme, and before the discovery of the new facts indicating a progressive problemshift in the latter, the objective relevance of the Meissner—Ochsenfeld experiment was very limited.

The strategy to be followed for the microscopic treatment of superconductivity was implicitly articulated through London's work, and by the fundamental changes brought about in the concepts used in electron theory: Superconductivity was not "simply" a quantum effect. It was a phenomenon in which correlated electronic motions seemed to exist over arbitrarily large macroscopic distances. Therefore, the search was for an attractive interaction between the electrons, and the problem had to be solved for many bodies, since the electron wave functions were *necessarily* correlated because of the long-range order in momentum space.

London's claim for such an order *specified* the particularity of the macroscopic quantum phenomenon. And after Fröhlich's (and Bardeen's) suggestion that the electron-lattice interaction could well be such an interaction, and his actual calculation with perturbation methods together with the almost simultaneous experimental corroboration of his prediction that the critical temperature was, in fact, sensitive to the isotope mass, the solution of the problem of superconductivity became in effect a mathematical problem. Summarizing, then, the salient features of the developments of the attempts to understand superconductivity, we have the following:

First, the truly remarkable discovery of the zero resistance of mercury at about 4 °K did not, strictly speaking, establish a paradoxical situation since it was conceivable — in the absence of a theory of electrical conduction in low temperatures — to incorporate such behavior within one of the existing models. The scientific community insisted for a long time on considering superconductivity as a case of perfect conductivity (thus regarding a referential property as a constitutive property).

Second, the belief in the non-diamagnetic character of superconductivity was due to faith in the correctness of the Maxwell equations and to the methodological attitude that such a "straightforward" derivation need not be experimentally tested. It was within such an overall framework that the (incorrect) interpretation of Kamerlingh Onnes' and Tuyn's experiment was believed to support the derivation of the non-diamagnetic character of superconductors. Thus, the subsequent discovery of the diamagnetic character of superconductors, even though it seemed to be violating a specific prediction, in effect obliged us to rethink our extrapolating procedures.

Third, despite the failures of the repeated attempts to resolve the problem of superconductivity, these were not (rational) violations of the constraints of the originally formulated (wrong) problem. Indulging now in a little fantasizing about "how things could have happened", violating constraints in this particular situation would have implied acceptance of the thermodynamic arguments; that, it is conceivable, could have led to the realization that considering superconductivity as a case of perfect conductivity was not the right way of approaching this peculiar phenomenon. If thermodynamics was to have been used for the explanation of superconductivity, and if, then, the scientific community tried to find a justification for the successful phenomenological solution (as happened later with superfluidity and the two-fluid model), then the discovery of the Meissner effect would have provided the appropriate justification.

Things did not happen that way, and we are, of course, not suggesting that this was the path they should have followed. We are simply pointing out to the limitations of a particular theoretical framework and the behaviour of a community trapped in it exactly because of the amazing successes of the very same framework.

5.2. Ordering and reordering

One can use the classical notions together with the classical descriptive language to be able to communicate, in a relatively satisfactory manner, the behaviour of the superconductors. This remains basically so even in light of the changes in various classical concepts which have been brought about by the development of the theory of superconductivity. The case with superfluidity, however, was radically different. Nearly all attempts to provide an explanation for the peculiar properties of helium expressed at the same time the inadequacy of the classical descriptive language as a means for communicating the processes responsible for these properties.

One may be tempted to claim that our proposed schema and the way the physical problems were formulated can be quite satisfactority used by first considering the paradoxical situation created by the "viscosity paradox" itself, and, then, by regarding the problem of superfluidity as a problem whose solution should provide an explanation to the "viscosity paradox". Such an approach is indeed correct, but tells only part of the story. It surely does not do any justice to the various developments which preceded the discoveries of 1938, which, in retrospect, seem to have been "necessary" for formulating the "problem of superfluidity".

Before the discovery of the discontinuities in the various thermodynamic parameters of liquid helium at T_{λ}, there was the discovery of a very peculiar property unique to liquid helium: Liquid helium remained liquid under its own pressure down to absolute zero. This highly "abnormal" situation was an indication — as Simon first pointed out in 1923 — that quantum characteristics were, in fact, decisive in determining the very *state of being* of this substance. It was not "simply" the case that we were observing quantum effects known to be present in all physical phenomena. In the case of helium, it was the *absence* of a state of being — the solid phase — that was due to the quantum behavior of the atoms involved and, thus, manifested characteristics of quantum physics on a macroscopic level.

It was not until 1923 that we had a direct indication of the existence of zero point energy. Keesom solidified helium under pressure in 1926 and deduced that the entropy difference between liquid and solid helium was tending to zero as the temperature was lowered. Was such a state of affairs a paradoxical situation, much in the same sense as the paradoxical situations described in Part I? If we do have a paradoxical situation, should it not, according to our schema, lead to the formulation of the right physical problem? Why, then, is superfluidity usually coined only with the problem associated with the "viscosity paradox"?

It is, we feel, misleading to hold that since helium is superfluid below T_{λ}, then all its "peculiar" properties are due to this state of helium and that their understanding would depend solely on the solution of the "problem of superfluidity". It is absolutely crucial to realize that what characterizes the development of the various explanatory schemata for superfluidity is that the community of researchers were confronted with *two* distinct physical problems and that the *formulation of the first* (and, to a certain extent, its resolution) became a prerequisite for being able to *formulate the second.* Most importantly, after having resolved the two physical problems, one can "re-read" the whole development of superfluidity and consider it as a successful attempt to elucidate and "resolve" what seems to be predominantly a conceptual problem.

Let us be more analytical:

The fact that helium is liquid under its own pressure down to absolute zero with the entropy difference between the liquid phase and the solid phase diminishing as the temperature decreases created a paradoxical situation. Since helium was solidified under pressure, it had a unique property: At low temperatures the internal energy of liquid helium was smaller than that of solid helium which was in equilibrium with it. This is a paradoxical situation in exactly the same manner as described in Part I. There was an incompatibility between the implications of the theoretical framework of classical kinetic theory and the implications of the description of such a phenomenon — a description which used almost exclusively the notion of order (or disorder). In this situation we did not really have a problem in superfluidity as such. Instead, we had a problem in thermodynamics which forced us to reformulate basic notions and mechanisms of molecular physics (as to a certain extent was, in fact, the case with the "viscosity paradox"). For if the main difference between a liquid and solid phase is that the former was less ordered and the latter, as in the case of helium, had greater internal energy, then this state of affairs could not be reconciled with the third law of thermodynamics. Let us note that there appeared a paradoxical situation at a very fundamental level where definitions, basic principles, and experimental results did not seem to make a coherent whole.

The problem was eventually resolved, and a new notion of order was established through a process of "concepts out o context(s)": The third law of thermodynamics was reformulated; the possibility of accepting an orderly state as a state in which the occupation of "places" was (quantum) probabilistically determined emerged; the notion of condensation of a Bose—Einstein aggregate was also reformulated. The work of Simon, London, and Fröhlich gave rise to a new theoretical framework TF_n, and it became possible to define an ordered state at zero momentum not in "real" space but, because of the uncertainty principle, in momentum space.

In any liquid there is a lattice-like arrangement of its molecules, and this usually gives the impression of a crystalline structure which, in turn, gives rise to observed X-ray diffraction rings. This is, however, a "local" property of the liquid — "there is [in the liquids] a certain degree of local order, without long-range order"[9] Initially, then, success was achieved in extending what was a locally valid property in nearly all the liquids to the whole of the liquid itself. As we have already mentioned, Fritz London, by assuming a specific kind of crystalline structure, was able to show that at absolute zero the total energy of liquid helium does, in fact, lie lower than that of solid helium. It was through London's work, wherein it became possible to associate a notion of order to *liquid helium as a whole*, that we have the first major conceptual changes in the development of the theories of superfluidity. The notion of order was redefined to allow us to *look* at what was considered (with the old notion) less ordered and to *see* it (with the redefined notion) in an ordered state.

London's work in superconductivity was crucial in helping him extend some fundamental ideas to understanding the properties of liquid helium. As a result of London's contributions to the development of theories of superconductivity, there was a truly remarkable metamorphosis to the concept of order. It was now possible to think of the various processes not in ("real") coordinate space, but rather in momentum space. "The electrons prefer to "solidify" with respect to their momenta rather than with respect to their coordinates."[10] This new notion of long-range order was not in coordinate space, but in momentum space.

It was such considerations which led London to think seriously about the possibilities implied by the results of Einstein's 1924 paper on the condensation of an ideal gas and to bring attention to this paper. Despite the fact that Einstein's conclusion was considered an oddity, and not much credence was given to its results concerning condensation, London was able to show that the specific condensation represents a phase transition of third order. It is to London's credit that, through a process similar to "concepts out of context(s)", he was able to provide a consistent (at least qualitative) explanation of the λ-phenomena by "stretching" the various concepts, which were originally used for the gaseous state, and applying them to the liquid state.

Starting from a theoretical framework wherein it was proved that an ideal gas showed this peculiar behavior of condensation — which could also be considered a *third* order transition — London was able to determine a new theoretical framework where these properties became meaningful for a liquid going through *second* order transitions. Interestingly, one now had a notion of condensation which had nothing to do with crystallization.

> It [helium II] actually is the one and only example of this phenomenon [Bose—Einstein condensation], the apparent complete order at very low temperatures is a regularity in the distribution of velocities, and not in the position of atoms. *Speaking in more mathematical terms, the condensation and the ordering takes place in "momentum space" rather that in coordinate space and the paradox inherent in the idea of an ordered fluid is removed.*[11] (Emphasis added.)

London's proposal was all the more remarkable if one takes into consideration that the proposed mechanism of Bose—Einstein condensation, and the ensuing changes in the notion of order, happened after a sweeping criticism of Fröhlich's proposal — a proposal which, as we remarked, considered the λ-point transitions as an order-disorder transition by reinterpreting the *already* reinterpreted concept of order which London himself had proposed for the successful explanation of the existence of liquid helium down to 0 °K.

Once it became possible to talk about order in momentum space, then the excess meaning of such a reinterpreted notion of order led quite naturally to determining the conditions under which one could talk about macroscopic quantum states. According to London, once we have such a notion of order, then "a dissolution of the system into separated wave packets of different molecules becomes impossible and that only a description by a common wave function for the whole system has a well-defined meaning."[12] Therefore, the macroscopic quantum character which we consider a unique feature of both superconductivity and superfluidity had been shown to be derivable from this redetermined notion of order.

The successful resolution of this problem and the proposed redefinition of the notion of order, coincided with the observation of the "viscosity paradox" and the fountain effect, and the publication of Tisza's two-fluid model. It was immediately realized that there was a model whose *success* as a phenomenological explanatory schema was all the more peculiar since it was based on a hypothesis which could not be justified within the dominant theoretical framework. The two-fluid model assumed that it was *possible to distinguish* two liquids made up of similar constituents and under the same conditions — something which no classical (and at that time not even quantum) framework permitted.

We will hold that what constituted a paradoxical situation was the very *success* of the two-fluid model which led to the formulation of the second problem mentioned above. The success of the two-fluid model warranted a microscopic justification and a reconsideration of the notion of motion since it was obvious that no satisfactory microscopic basis could be provided unless it also involved a redefinition of the concept of motion. Therefore, understanding the success of the two-fluid model begged for a "double" explanation. First the theoretical justification for the success of the model was needed. Secondly, any solution would have *necessarily* entailed a new interpretation not readily provided by the existing theoretical framework.

Landau's monumental contribution was to develop a theoretical framework for reinterpreting what were considered to be different fluids in classical language as different quantum *states*. Let us quote a very characteristic passage from his paper of 1941 where he developed his quantum hydrodynamics.

It must be stressed that when we look upon helium as a mixture of two liquids

> it is no more than *a method of expression convenient for describing phenomena which take place in helium II. Like every description of quantum phenomena in classical terms it is not quite adequate.* Actually one must say that in a quantum liquid two movements can exist simultaneously, each of which is connected with its own "effective mass" (so that the sum of both these masses equals the total real mass of the liquid). One of these movements is normal (i.e., it possesses the same properties as the movements of the usual liquids); the other is superfluid. Both of these motions take place without transfer of the momentum from one to the other. We emphasize in particular that there is no division of the real particles of the liquid into "superfluid" and "normal" ones. In a *certain sense* one can speak of "superfluid" and "normal" masses of liquid as masses connected with the two simultaneously possible movements, but this *by no means signifies the possibility of a real division of the liquid into two parts.*
>
> Keeping in mind these reservations as to the real nature of the phenomena taking place in helium II, it is still convenient to use the terms "superfluids" and "normal" liquids as a short and convenient method of describing these phenomena[13] (Emphasis added).

Landau fully realized the serious problems involved in the interpretation of this newly (re)formulated concept of motion, and he was extremely careful while introducing these concepts to point out that they should not be taken "literally". It is obvious from his writings that Landau was not only striving to develop a new theory, but wanted also to develop a different descriptive language parallel to the development of the new theory.

Even after the attempt to propose a microscopic theory, the question as to whether Bose—Einstein condensation was an essential feature for understanding superfluidity was justifiably posed, and remained unanswered. This does not lead to the conclusion that Landau's theory did not provide the needed basis for the two-fluid model. It is rather an argument *in favour* of our claim, since, if the problem to be solved was the theoretical justification of the two-fluid model, then there will *not* be a single way to reach this justification. Whether quantum statistics, and especially Bose—Einstein condensation, is necessary for comprehending superfluidity is not something which had to be answered on a priori grounds (as, in fact, the subsequent experimental developments with ^{3}He showed). The question itself was meaningful only because of London's original theory.

Even after Landau's theory, London continued to talk about the need for a microscopic theory of superfluidity and, therefore, for the theoretical justification of the two-fluid model whose successes were reviewed and continued to be impressive in 1946. London's criticism of Landau's theory was extremely significant methodologically at two levels. He maintained that Landau's theory did not really predict anything new as did Tisza's proposal. Therefore, it was implied, at best, it would be only a justification of the two-fluid model and not a separate, let alone antagonistic, theory to the two-fluid model. London, in 1946, maintained that the two-fluid model, based at it was on the idea of Bose—Einstein condensation, did in fact predict new effects which were subsequently observed experimentally. London — somewhat unjustifiably — reserved for Landau's theory an indictment deserved only on methodological grounds. "Landau's theory appeared too late to make all these inferences in form of predictions!"[14]

But London himself was quite inconsistent with the way he had chosen to assess the relative merits of his and Tisza's approach and that of Landau's: in 1946 at Cambridge, almost all his emphasis was on the second sound measurements and hardly mentioning the question of He^3. The latter became crucial in his arguments after the second sound measurements below 1 °K appeared to corroborate Landau's predictions.

London's second critism had to do with the ontological status of the "rotons". "Landau presented his theory as if it were an outcome of his quantum hydrodynamics, which in itself is a very interesting attempt. The quantization of the motion of a liquid, as he proposed it, is, however, quite unconvincing as far as it is based on a representation of the states of the liquid by phonons and what he calls "rotons". There is unfortunately no indication that there exists anything like a "roton"; at least one searches in vain for the definition of the word. It should not have been difficult to give such a definition if there were as clear and simple a concept as, for example, the one we know for the phonons of the solid body. There is, further, no trace of a proof that these phonons and rotons, if the latter were to exist, would give rise to system of two interpretrating fluids"[15].

When Landau first proposed the rotons he was very careful not to give them a "real status"; he was very concerned about "how" real they were. His writings definitely give the impression that he was fully aware of the difficulties concerning the overall assessment of his theory because of the ambiguous status of the rotons. This is one of the reasons that when in 1948 Landau answered Tisza's criticisms he chose to go back and defend the unquestionably quantum mechanical notion of "elementary excitations", and not the "rotons" which all along he considered a special type of these excitations. He firmly asserted that "it follows unambiguously from quantum mechanics that for every slightly excited macroscopic system a conception can be introduced of "elementary excitations" which describe the "collective" motion of the particles . . . It is this assumption, indisputable in my opinion, which is the basis of the microscopic part of my theory. On the contrary, every consideration of the motion of individual atoms in the system of strongly interacting particles is in contradiction with the first principles of quantum mechanics".[16]

Despite the fact that London's criticisms appeared to be questioning the validity of Landau's theory, it was Landau who was "conceptually" on the right track, and it was eventually the "rotons" which, when reinterpreted, were to be given a physically meaningful status. London had based his criticism on the intuitive character of Landau's initial assumptions, on his belief that there could not be a justifiable meaning of the roton within the "hierarchical conceptual structure" of the TF_i, and, of course, on the fact that Landau's derivation of superfluidity did not depend on the specific statistics of the helium atoms. On the other hand, Landau insisted on giving his concepts contextual meaning, which of course, could not be adequately given unless there was a change in the respective theoretical framework.

This was the contribution of Bogoliubov and Feynman. Both physicists, before tackling the problem of superfluidity, had made important contributions to quantum field theory, and they extended the techniques developed there to liquid helium. Bogoliubov was able to derive Landau's assumption of elementary excitations from the equations expressing the behaviour of a non-ideal Bose–Einstein gas. Feynman, using a formalism he had already developed for quantum electrodynamics, was able to propose an explanation of the λ-point transitions. What is important from our point of view is that both theories unified the London-Landau approaches, without changing the basic viewpoint of each theory. The rotons were now likened to "a kind of quantum mechanical analog of a microscopic vortex ring".[17]

Even though most of the unusual properties of helium were first observed at Leiden, it is common practice to hold that "proper" work for understanding the peculiar behavior of helium started after the dramatic experimental developments were reported in *Nature* in 1938. The situation is sometimes likened to superconductivity in that it was only after the discussion of the viscosity paradox and the fountain effect that superfluidity (much like superconductivity after the discovery of the Meissner effect) could be properly treated.

In the case of superconductivity the observation of zero resistance could not be used to formulate the right physical problem especially when people thought that superconductors were nothing but perfect conductors, and not diamagnetic, at low temperatures. In the same manner, the property, for example, of super heat conductivity ("the exact analog to infinite conductivity")[18] could not be adequately used to formulate the problem of superfluidity, since people continued to think that the standard tests for determining the viscosity of a liquid should all yield the same results. Furthermore, *in the same way that the discovery of the Meissner effect led to the formulation of the right physical problem by redefining the differences between a superconductor and a perfect conductor, the discovery of different values for the viscosity of liquid helium by the "standard" tests led to the formulation of the problem of superfluidity.*

When superconductivity was discovered, there was hardly any reliable theory for electrical conduction; for low temperatures, there was no prediction of a unique result. Hence, technically speaking, what was unexpected in superconductivity was the *sudden* drop of resistance at temperatures above absolute zero. On the other hand, the period beginning with the late twenties, when we have the first indications for the abnormally high heat conductivity of liquid helium below T_λ, and ending in the thirties, when this property was firmly established, brought the development of the quantum mechanical electron theory. The theory provided specific predictions for heat conductivity at low temperatures, i.e., that it should diminish as the temperature is lowered. Therefore, in the case of the infinite electrical conductivity, a different result was obtained from what was predicted by one of the models. In the case of the abnormally high heat conductivity with correspondingly small temperature gradient, one was confronted with a *falsification* of a specific prediction of a well-formulated theory.

However, apart from the difference between these two seemingly similar cases, there is an implied assumption, which, despite its appeal in promising a unified view for both superconductivity and superfluidity, is itself self-defeating. There is a limit to how much we can generalize if we treat both the electrical current and the heat current as generalized "flows". Notwithstanding the success of Lord Kelvin and J. C. Maxwell in developing electromagnetic field theory by making use of mathematical analogies with theories of fluid flow and heat conduction, the notion of "flow" is primarily used to denote a mechanism in a *generic manner*; i.e., where the output will have to be quantitatively compared to the input which is usually also of the same quality as the output.

If the notion of flow in the classical framework is extremely useful for comprehending a large class of "input-output" phenomena, it is equally misleading to use it for understanding quantum phenomena — especially when the two macroscopic quantum phenomena can by no means be treated as "input-output" phenomena. Interestingly, these two phenomena resisted even phenomenological "explanations" so long as they were treated as "input-output" phenomena.

We obviously could have "read" the development of superfluidity theory by starting with a paradoxical situation created by a new and unexpected phenomenon (either the superheat conductivity or the viscosity paradox). Even though that would have been a starting point consistent with our proposed schema, it should be mentioned that our schema can accommodate equally the development of cases where the original problem is the result of a paradoxical situation that is not between the implication of the description of a new and unexpected phenomenon, and the implications of the theoretical framework of the dominant model or theory, but rather between the implications of TF_i and those of a newly-proposed model which gives a satisfactory account of the observed phenomena. A paradoxical situation may also arise when the very success of a model is due to assumptions which are incompatible with the $J(TF_i)$.

Summarizing, the most salient features of superfluidity, we have the following.

Understanding the phenomenon of superfluidity was not a matter of resolving a single physical problem, but rather of being able to formulate at least three "right problems". First, a paradoxical situation arose after helium was observed to remain liquid down to absolute zero under its own pressure. Second, there was a paradoxical situation when the λ-point transition and the discontinuity of the thermodynamic parameters did not turn out to be an expression of a (normal) change of phase. Third, the success of the two-fluid model with its highly unphysical assumptions could not be readily accounted for by the dominant theory. Strictly speaking, the "problem" was not really the success of the two-fluid model. It was, rather, the fact that its success warranted both a theoretical justification on a microscopic level together with a reconsideration of the notion of motion. The resolution of all these problems brought about radical changes in the meaning of various classical concepts, first and foremost in that of "order". It is possible to "read" the whole development of the theories proposed to explain the phenomenon of superfluidity as attempts to resolve a conceptual problem related to the reinterpretation of the concept of order.

The reinterpretation of the concept of (molecular) motion followed from an analogous successful reinterpretation of the concepts of order and disorder and the fact that it became possible to view what was considered a classically disordered state as a quantum mechanical ordered one.

Notes

Chapter 1

1. N. R. Hanson, 1958b.
2. T. S. Kuhn, 1962a.
3. N. R. Hanson, 1958b, p. 633.
4. A detailed discussion of these is found in Section 1.5.
5. K. Gavroglu, 1985.
6. The use of the term falsification does not mean that we subscribe to the Popperian approach concerning the testing of theories. It signifies those phenomena that the scientific community is obliged to explain satisfactorily with any proposed new theory.
7. K. Gavroglu, Y. Goudaroulis, 1985.
8. P. Feyerabend, 1975, p. 68.
9. R. Dicke, 1961. For a methodological analysis of Dicke's approach to the General Theory of Relativity see K. Gavroglu, 1986.
10. We emphasize that a paradoxical situation is *not* a situation created because of a hitherto unknown contradiction either in the mathematical structure of the theory or in its physical assumptions.
11. Nancy Cartwright's (1983) claims do give a special status to the phenomenological laws. Her criticism of the realist interpretation of physics relies on emphasising the argument that the phenomenological laws as they are used in physics do, in fact, describe the actual situation to which they are applied. And despite the fact that "the great explanatory and predictive powers of our theories lies in their fundamental laws" Cartiwright proposes that the "*content* of our scientific knowledge is expressed in the phenomenological laws" . . . Once a situation is adequately described by them then this situation is real".

    However, in order to argue such an overall approach one has to provide specific answers to at least the following questions: how do we decide that an account of a phenomenon is a phenomenological law, and not just a "fancier" expression of the description of the phenomenon? In other words, what are the criteria for accepting a schema as a phenomenological law which "transcends" the descriptive account of the phenomenon? What are, during each period, the specific features of the generally neglected historical character of the relationship between phenomenological and fundamental laws? These questions do not seem to be answered in a satisfactory manner.

    Aesthetic reasons and economy of parameters have always been decisive factors both in proposing phenomenological laws and in the discussions concerning their acceptability by the scientific community. Often, however, the phenomenological laws imply or explicitly state new and novel "mechanisms". And there is not an insignificant number of cases where the proposed mechanisms are "unusual". In such cases the effort is not only to derive these laws from more fundamental ones, but, also, by investigating the possibilities provided by the more fundamental laws to give credence to these "unusual mechanisms". This is not necessarily achieved by proposing an equally unusual mechanism at a more fundamental level and shifting the "unphysicalness" towards a more fundamental level since, then, it becomes more acceptable! Giving credence to such phenomenological laws is primarily a conceptual task whereby the introduction of new concepts and the reinterpretation of the old ones form a conceptually coherent context which facilitates the acceptability of the "peculiar mechanisms" by the scientific community.

12. Such is the situation in present day high energy physics where there are no outstanding experimental anomalies, and theoreticians are mostly preoccupied with devising unifying schemata and theoretically (or is it methodologically and metaphysically?) more satisfying explanations for the existing data. These schemata often involve mechanisms which cannot be "justified" within the presently held interpretation of the theoretical framework of relativistic quantum physics. The situation has many similarities with the case of the two-fluid model in superfluidity that we will discuss analytically in Chapter 4.
13. T. Nickles, 1978, 1980b, 1980c, 1980d.
14. K. Popper, 1972.
15. *Ibid.*, p. 170.
16. *Ibid.*, p. 181.
17. *Ibid.*, p. 177.
18. Popper, unfortunately, does not systematically pursue his ideas on the "problem situation". Eventually he talks almost exclusively about problems.
19. *Ibid.*, p. 181.
20. *Ibid.*, p. 181.
21. T. Nickles, 1978, p. 10.
22. T. Nickles, (forthcoming).
23. I. Lakatos, "Falsification and the Methodology of scientific research programmes", in I. Lakatos, 1978, Vol. I, pp. 8—101.
24. See especially A. Musgrave, 1976, and R. S. Cohen et al., 1976.
25. J. Watkins, 1970, p. 31.
26. See K. Gavroglu, 1989.
27. I. Lakatos, note 24, p. 51.
28. *Ibid.*, pp. 57—58.
29. *Ibid.*, p. 50.
30. *Ibid.*, p. 51.
31. S. Weinberg, 1980, p. 517.
32. I. Lakatos, note 24, p. 88.
33. L. Laudan, 1977.
34. *Ibid.*, p. 68.
35. S. Weinberg, 1980.
36. L. Laudan *et al.*, 1986, p. 208.
37. A. Baltas, 1986.
38. I. Hacking, 1983; D. Gooding *et al.*, 1988; A. Franklin, 1986. P. Gallison, 1987.
39. I. Lakatos, note 24, p. 77, footnote 2.
40. E. Zahar, 1973, points out that the re-definition of the novel fact implies much more than the already revitalized attention to the totality of the questions related with the *historical* research on the origins and developments of the various theories, *and thus*, the relationship between the theories and the empirical data within a research program is not so straightforward.
41. The best systematic treatment of the notion of approximation in theories is to be found in C. V. Moulines, 1976.
42. For a different use of "concept innovation" see J. Sneed, 1986.
43. Even this limit is highly unrealistic and it involves the "best possible" developments.
44. D. Gross, 1986.

Chapter 2

1. One should note, however, that there was no possibility of preparing liquid helium at Leiden during World War I, and experiments were resumed in 1919 when the U.S. Government gave Onnes 30 m^3 helium gas.
2. E. Cohen, 1927, J. F. H. Koopman, 1927. Also see *Scientific American* February 13, 1915. Heike Kamerlingh Onnes was born at Groningen on September 21, 1853 and entered the University of Groningen to study physics and mathematics where, at the end of his first year in 1871, he won a gold medal awarded by Utrecht University with his essay titled "A critical investigation of the methods for determining vapour density and of the results obtained thereby, with respect to the relation of the nature of the chemical compounds and the density of their vapours". He went to Heidelberg at the end of 1871 where Bunsen and Kirchoff were professors, and where extensive work was being done in spectrum analysis.

   In 1872, he received the second prize at a competition initiated by the Senate of the University of Groningen with his essay titled "A critical survey of the methods for determining the quantities of heat which are set free by chemical reactions and dissociations and of the results obtained by different investigators".

   In 1873 he returned to Groningen and, in 1879, he received his doctor's degree from the same University. His thesis, "New proofs for the axial rotation of the Earth", had been the continuation of work he had already started with Kirchoff at Heidelberg and in it he displays an impressive command of mathematics. A year before receiving his Ph.D. he had been appointed assistant to Bosscha's laboratory at the Polytechnic School (and, later, Technical University) at Delft, while being in close contact with J. van der Waals, who was then at Amsterdam.

   Before Kamerlingh Onnes' appointment at Leiden two Dutch physicists who were to make really fundamental contributions to physics, had already received their doctoral theses from Leiden: J. van der Waals in 1873 completed his dissertation on the continuity of the gaseous and liquid states, and in 1875 H. Lorentz completed his dissertation on the reflection and refraction of light, making extensive use of Maxwell's equations. Kamerlingh Onnes' research programs (as we will show analytically) were strongly influenced by the work of both van der Waals and Lorentz and Onnes's work in molecular physics as well as on the magnetic electric and optical properties of various substances was an outgrowth of the work of these physicists. Their remarkable theoretical contributions inspired a program in experimental physics whose realisation in such a relatively short time with results which were to have such far-reaching consequences, remains unique to our days.

   Even before the liquefaction of helium, Kamerlingh Onnes' work related to the liquefying techniques, and especially his contribution to thermometry, had already been mentioned in various textbooks. His objections to De Heen's experiments in 1896, which seemed to disprove the continuity of the liquid and gaseous states as proposed by van der Waals and the ensuing discussion (which eventually justified Onnes) made him quite well known to the French speaking physicists.

   But even after the liquefaction of helium and the discovery of superconductivity, Onnes was not so well known, and it was only after receiving the Nobel prize in 1913 that he and his work became widely known: it is characteristic that *Scientific American* ran the first article on him in 1914.

   Kamerlingh Onnes, in 1885, founded the *Communications from the Physical Laboratory at the University of Leiden*, where all papers with results from experiments performed in the Laboratory were published in English with very few in French and German. The papers were originally published in Dutch in the *Proceedings of the Royal Academy at Amsterdam* and later in *Physica*. The *Communications* remain a remarkable chronicle of the work done in the Leiden Laboratory and of the development of low temperature physics up to about the end of the twenties.

   One should also mention that, in 1901 he founded the "Society for the Promotion of the

Training of Instrument Makers" in the form of a training school annexed to the University of Leiden. He was also one of the persons actively involved in the founding of the "Association Internationale du Froid" in 1909.

3. E. Cohen, 1927, p. 1197.
4. An excellent presentation of some aspects of the developments in instrumentation can be found in R. de Bruyn Ouboter, 1986.
5. J. D. van der Waals, 1880; 1888; 1910; 1912; J. de Boer 1974; M. J. Klein 1974.
6. H. Kamerlingh Onnes, 1882.
7. J. D. van der Waals, 1888, p. 454.
8. H. Kamerlingh Onnes, 1896, p. 5.
9. E. Cohen, 1927, p. 1203.
10. H. Frohlich, 1961.
11. K. Gavroglu, Y. Goudaroulis, 1984; P. F. Dahl, 1984, 1986.
12. H. Kamerlingh Onnes, 1882.
13. Characteristically we mention two important papers by one of the main researchers in the field M. E. H. Amagat, 1896a, b. See Mathias, 1894, who does not include Onnes' contribution among the people he mentions who proposed a generalization of the law of corresponding states. And also J. M. H. Levelt Sengers 1974.
14. J. D. van der Waals, 1890, p. 43.
15. H. Kamerlingh Onnes, 1894, p. 4.
16. H. Kamerlingh Onnes, 1896, p. 5.
17. H. Kamerlingh Onnes, 1901a, b.
18. H. Kamerlingh Onnes, 1901a, p. 3.
19. H. Kamerlingh Onnes, 1904, pp. 18—19.
20. H. Kamerlingh Onnes, C. Zakrzewski, 1904, p. 3.
21. W. Keesom, 1901.
22. There are 35 publications between 1901—1928 on the isotherms of the diatomic gases and their binary mixtures, and 26 publications between 1907—1927 on the isotherms of the monoatomic gases and their binary mixtures.
23. H. Kamerlingh Onnes, H. H. Francis Hyndman, 1901, 1902; W. J. de Haas, 1911, 1912.
24. H. Kamerlingh Onnes, W. J. de Haas, 1912, F. P. G. A. J. van Agt, 1925.
25. H. Kamerlingh Onnes, 1917.
26. H. Kamerlingh Onnes, C. A. Crommelin, 1913.
27. F. P. G. A. J. van Agt, 1925.
28. H. Kamerlingh Onnes, W. Keesom, 1912.
29. See mainly H. Kamerlingh Onnes, H. A. Kuypers, 1924.
30. H. Kamerlingh Onnes, C. Dorsman, G. Holst, 1914.
31. J. Palacios Martinez, H. Kamerlingh Onnes, 1923.
32. H. Kamerlingh Onnes, W. Keesom, 1908.
33. H. Kamerlingh Onnes, Cromelin, C. A., 1911.
34. H. Kamerlingh Onnes, W. Keesom, 1912.
35. W. Keesom, 1912a, b, c, d and 1915a, b.
    W. Keesom, Miss C. van Leeuwen, 1915.
36. E. Mathias, 1928.
37. E. Cohen, 1927, p. 1203.
38. The magnetic researches of Kamerlingh Onnes involved the study of the magnetisation of liquid and solid oxygen and its compounds; the measurements of the susceptibility of nickel, oxygen, solid oxygen and of mixtures of liquid nitrogen and oxygen, and gadolinium sulphate; the study of paramagnetism and diamagnetism at liquid hydrogen temperatures; and the investigations of the magnetic properties of the paramagnetic double sulphates, chromium chloride, paramagnetic chlorides and gadolinium sulphate.
39. M. P. Curie, 1895.
40. P. Weiss, H. Kamerlingh Onnes, 1910.
41. H. Kamerlingh Onnes, A. Perrier, 1911a, b.

42. H. Kamerlingh Onnes, E. Oosterhuis, 1912, 1913a.
43. A. Perrier, H. Kamerlingh Onnes, 1914b.
44. H. Kamerlingh Onnes, 1914d.
45. G. Breit, H. Kamerlingh Onnes, 1923.
46. H. Kamerlingh Onnes, A. Perrier, 1910.
47. H. Kamerlingh Onnes, A. Perrier, 1911a, b.
48. H. Kamerlingh Onnes, A. Perrier, 1911a.
49. *Ibid.*
50. P. Weiss, 1907, 1910.
51. H. Kamerlingh Onnes, E. Oosterhuis, 1913a, p. 52.
52. H. Kamerlingh Onnes, E. Oosterhuis, 1913b.
53. A. Perrier, H. Kamerlingh Onnes, 1914b.
54. *Ibid.*, p. 39.
55. *Ibid.*, p. 40.
56. *Ibid.*, pp. 52—53.

Chapter 3

1. J. Dewar, J. A. Fleming, 1893, (p. 296). See also J. Dewar, J. A. Fleming, 1896, (p. 81): "These measurements, therefore, afford a further confirmation of the law we have enunciated as a deduction from experimental observations, that the electrical resistivity of a pure metal vanishes at the absolute zero of temperature".
2. A. M. Clerke, 1901, (p. 705).
3. The assumption which J. J. Thomson could not avoid in an effort to explain the Thomson effect is that $n \propto \sqrt{T}$ and since $v \propto \sqrt{T}$ we must have $\lambda \propto 1/T$.
4. H. Kamerlingh Onnes, J. Clay, 1906a, (p. 39).
5. H. Kamerlingh Onnes, J. Clay, 1907a, (p. 18). See also H. Kamerlingh Onnes, J. Clay, 1907b.
6. H. Kamerlingh Onnes, J. Clay, 1908, (p. 26).
7. H. Kamerlingh Onnes, 1911a. For a detailed presentation of the discovery of superconductivity and the work done during the first years following the discovery see P. F. Dahl, 1984, 1986.
8. H. Kamerlingh Onnes, 1913e, (p. 330).
9. H. Kamerlingh Onnes, 1913d, (p. 59). The electrical resistance of alloys does not, in general, become very small at low temperatures and this suggested that the purity of the metals is of great importance for temperature—resistance investigations.
10. H. Kamerlingh Onnes, 1911a.
11. H. Kamerlingh Onnes, 1911e, (p. 13).
12. H. Kamerlingh Onnes, 1911f, (p. 24).
13. H. Kamerlingh Onnes, 1913a, b; p. 3 and p. 35.
14. H. Kamerlingh Onnes, 1913b, (p. 31). This may be a reason why the otherwise careful E. Cohen says that superconductivity was discovered in 1913.
15. H. Kamerlingh Onnes, 1913e, (p. 330).
16. H. Kamerlingh Onnes, 1913d, (pp. 58—59).
17. *Ibid.*, (pp. 59—60).
18. W. Nernst, 1911, (p. 313). F. A. Lindemann, 1911.
19. H. Kamerlingh Onnes, 1913d, (p. 62).
20. H. Kamerlingh Onnes, 1912, (p. 10).
21. In 1914 Kamerlingh Onnes' investigations on this matter permitted him to conclude that "with respect to the specific heat nothing peculiar happens at the point of discontinuity" and that "there is no reason for speaking of an allotropic modification at the discontinuity point of conductivity". These experiments, together with some theoretical speculations, had led to some then generally accepted ideas: one assumed that there was no phase transition and

tried to explain a sudden change of the free path of the electron. [See H. Kamerlingh Onnes, 1914e, (p. 30); H. Kamerlingh Onnes, 1921, (p. 45)].

Some years later, when the technique of calorimetric measurements at these very low temperatures had been much improved and one could perform these measurements much more accurately than before, Keesom and his collaborators found that the atomic heats of lead, tin and zinc undergo a rapid change and concluded that "the rapid change or jump is connected with superconductivity" [W. H. Keesom and J. N. van der Ende, 1932, (p. 10)].

22. H. Kamerlingh Onnes, 1912, (p. 10).
23. H. Kamerlingh Onnes, 1913b, (pp. 43—44).
24. *Ibid.*, (p. 44).
25. See *Ibid.*, (p. 46).
26. H. Kamerlingh Onnes, 1913a.
27. H. Kamerlingh Onnes, 1911e. See also H. Kamerlingh Onnes 1913c, where he claimed that "The possibility of using the superconductors tin and lead, gives a new departure to the idea of Perrin of procuring a stronger magnetic field by use of coils without iron . . . even a coil of 25 cm diameter of lead wire . . . could give a field of 100 000 gauss, without perceptible heat being developed in the coil . . . this remark may serve to put the problem of very strong magnetic fields which are becoming indispensable for various investigations in a new form". [*Ibid.*, (p. 65) and (fn. 1 in p. 64)]. After the discovery of superconductivity of mercury in 1911, Kamerlingh Onnes made extensive measurements to establish new superconductors. Pyrite did not show such a behaviour, and tin and lead did. Superconductivity was also established in thallium, indium, sodium, and potasium. Further measurements of the resistance of mercury, tin, cadmium, constantin, manganin and copper were undertaken in order to find suitable resistances for thermometers.
28. H. Kamerlingh Onnes, 1914b. P. Ehrenfest described to Lorentz his excitement after seeing one of Kamerlingh Onnes' experiments of superconductivity in which a current once established, continued to flow on its own for over 24 hours: "It is uncanny to see the influence of these 'permanent' currents on a magnetic needle. You can feel almost tangibly how the ring of electrons in the wire turns around, around, around-slowly, and almost without friction" [P. Ehrenfest to H. A. Lorentz, 11 April 1914, quoted in M. J. Klein, 1970, (p. 214)].
29. H. Kamerlingh Onnes, 1914b, (p. 11).
30. See J. J. Thomson, 1915, esp. pp. 192—3 and 198.
31. *Ibid.*, (p. 198).
32. Lindemann's view later developed by Borelius, 1918, and Haber, 1919.
33. H. Kamerlingh Onnes, 1921. Three years later Kamerlingh Onnes reiterated his views and his confidence in the Rutherford—Bohr atom and the quantum theory in a report to the fourth Solvay Congress in 1924. In this Congress Lorentz reviewed theoretical aspects of the entire subject, but threw little light on superconductivity beyond recalling the early attempts to rework classical theories by incorporating elements of quantum theory. Nor could Richardson add to the subject in his presentation of yet another theory of metallic conduction in which he threated superconductivity in terms of tangential Bohr orbits in solids. The Fermi statistics, essential for the understanding of metallic conduction, was still non-existent. [See H. Kamerlingh Onnes, 1924; H. A. Lorentz, 1924; M. O. W. Richardson, 1924]. In the same Congress Bridgman presented a rather detailed review of the theories of conductivity proposed after 1900. [See M. P. W. Bridgman, 1924].
34. H. Kamerlingh Onnes, 1921, (pp. 49—50).
35. H. Kamerlingh Onnes, B. Beckman, 1912a.
36. H. Kamerlingh Onnes, B. Beckman, 1912b.
37. H. Kamerlingh Onnes, 1913d, (p. 65, fn.1). Compare this to his earlier reaction when he failed (in 1910) to solidify helium: In his 1922 (p. 149), he emphasized not the failure but the novelty of the newly accessible temperature range. "This failure with regard to the solidification of helium meant a gain: a new region of temperature, which on account of its extreme situation is especially important, was proved to be accessible to us".
38. H. Kamerlingh Onnes, 1914a, (p. 71). One year before [see his 1913d, (p. 69)] Kamerlingh

Onnes had expressed this by means of an analogy: "If we make ourselves free from accepted theoretical ideas, we can draw attention to the analogy with the generation of waves on water when the velocity of the wind exceeds a certain value. There is perhaps even a deeper ground in this analogy pertaining to the question of stability of motion".

39. ". . . the phenomenon of threshold current need not be regarded as a distinct phenomenon, to be explained by heating, or otherwise, but as a direct result of the existence of the phenomenon of threshold magnetic field" [F. B. Silsbee, 1916, (p. 598)]. Silsbee concluded that "if it is true, as indicated in this paper, that the magnetic effect is the more fundamental, it would seem that this might afford a valuable clue leading toward a more satisfactory theory of the superconducting state and perhaps of metallic conduction in general".
40. W. Tuyn, H. Kamerlingh Onnes, 1926. The conclusions are expressed in Kamerlingh Onnes's characteristic way: ". . . on the faith of the results obtained up till now we think we may accept the hypothesis of Silsbee as being correct". [*Ibid.*, (p. 37)].
41. G. J. Sizoo, H. Kamerlingh Onnes, 1925, (p. 13).
42. G. J. Sizoo, W. J. De Haas, H. Kamerlingh Onnes, 1926, (p. 29).
43. See G. J. Sizoo, H. Kamerlingh Onnes, 1925, and G. J. Sizoo, W. J. De Haas, H. Kamerlingh Onnes, 1926.
44. H. Kamerlingh Onnes, W. Tuyn, 1922, (p. 13). However, ordinary lead and uranium lead were found to have the same T_c within the accuracy of 0.025 °K.
45. H. Kamerlingh Onnes could not have obtained a positive result within his experimental errors, since the respective atomic masses imply a difference in T_c of only 0.02 °K.
46. See, H. Kamerlingh Onnes, 1924, and W. Tuyn, 1929. Cf. K. Mendelssohn, 1964, (p. 8): "I think the reason is that none of us checked the details of the experiment with a lead sphere- which unfortunately was hollow- and we took this result for granted". Cf. also H. Casimir, 1983, (p. 339).
47. H. B. G. Casimir, 1977, (p. 170).
48. F. Bloch, 1980, (p. 27).
49. F. London, 1950, (p. 142).
50. See text to footnote 20.
51. See the discussion on H. Kamerlingh Onnes' (1924) report, presented by Keesom, esp. (pp. 285—289).
52. P. H. Van Laer, W. H. Keesom, 1938. See also C. J. Gorter, 1964, esp. (p. 4).
53. Its derivation by A. J. Rutgers was first drawn attention to in a note added in proof to P. Ehrenfest's, 1933. The formula was first explicitly derived in print by Rutgers, in his 1934.
54. C. J. Gorter, 1964, (p. 4).
55. W. J. de Haas, J. Voogd, 1931. See also W. J. de Haas, J. Voogd and J. M. Jonker, 1934.
56. H. B. G. Casimir, 1973, (p. 486).
57. H. B. G. Casimir, 1977, (p. 178).
58. C. J. Gorter, H. Casimir, 1934. See also B. S. Chandrasekhar, 1969, (pp. 24—25).
59. A general proof has been given by Von Laue. See: M. von Laue 1949, pp. 7f.
60. Because of the microscopic interpretation of λ this distance can be expected to be of the order of magnitude 10^{-6} cm. Actually, λ had been computed at least twice before (in 1925 by Mrs Lorentz-De Haas, and in 1933 by Becker, Heller and Sauter) but only with London did it take its accurate physical meaning. It is now usually called *London's penetration depth.*

    It is interesting to note here that in 1925, Mrs. De Haas — born Lorentz — published a note referring to the question as to how a variation of the external magnetic field could ever exert any influence on a superconductor if, according to Kamerlingh Onnes' view, it was completely screened off by a surface current. She pointed out that this so-called surface current should consist of n moving electrons per unit volume with charge e and mass m. In the usual reasoning on superconductivity, the kinetic energy $(1/2)mv^2$, where v is the average velocity of the electrons, was neglected in comparison with the electromagnetic energy. She computed at which size the two similar contributions would be equal and arrived at the statement that this would be at a few hundred Angstrom. [See G. L. De Haas-Lorentz, 1925. See also C. J. Gorter, 1964, (p. 4)].
61. F. London, H. London, 1935, p. 87.

62. The attraction between electrons increases the energy difference between the ground state of the electron gas and its first excited state. To remove one electron and to put it into a higher unoccupied level needs some definite energy, which is also called the 'gap'. An energy gap is believed to be a characteristic feature of the superconducting state. Some experiments on thermal properties gave evidence of it. There have been also several suggestions through the years for an energy gap on theoretical grounds. London's first suggestions was developed by H. Welker, 1938, in an attempt to account for the Meissner effect.
63. See L. Brillouin, 1935. See also F. London, 1935, (p. 25) and 1937, (p. 7). For a more formal proof of the theorem see M. R. Schafroth, 1960, (pp. 404—406).
64. F. London, 1935, (p. 31).
65. *Ibid.*, (p. 27).
66. This is true only for a simply connected superconductor and if the field is described in the gauge div **A** = 0. In multiply connected superconductors, this is not generally possible. A persistent current in a ring is metastable rather than stable. In this case, the flux through the ring is maintained at a constant value when the external field is changed.
67. F. London, 1950, (p. 150).
68. C. J. Gorter, 1964, (p. 6). Shortly after the first suggestion of F. London that the explanation of superconductivity depends essentially on quantum theory, Landau worked out a theory of the intermediate state as a series alternating lamina of normal and superconducting regions. His theory was confirmed shortly after the war in a series of experiments done in Moscow.

    At about the same time, in 1936 and 1937, Shubnikov discovered a new class of superconductors, which he called type II. In these, flux starts to penetrate at a lower critical field, which may be well below the thermodynamic critical field H_c, but the metal remains superconducting to an upper critical field $> H_c$. [See L. Landau, 1937b; 1938 and 1943. See also D. Shoenberg, 1952].
69. L. D. Landau, V. L. Ginzburg, 1950. See also V. L. Ginzburg, 1955. The Ginzburg-Landau theory was originally published in Russian in the days before such articles were routinely published in translated form, and "was largely ingored in the West until after 1960". [See D. M. Ginsberg, 1970, (pp. 949—950)]. See also J. Bardeen, 1973, (p. 43): "At the time this work was done, the Cold War was at its height. Distribution of Russian journals was slow. I was fortunate to obtain an excellent translation of the Ginzburg—Landau article through the courtesy of Shoenberg".

    On this point see also P. W. Anderson, 1969, (p. 1347): "1950 was in the midst of the McCarthy era, a time of which one of the silliest manifestations was the banning of Russian scientific publications in the United States. Some were even dumped into the harbors. The JETP containing the paper of Ginzbourg and Landau which, more even than the Western developments, marks the beginning of the modern era, was one of these. . . . There followed a tragicomic period of over a decade which should be fascinating to the historians of science and to those concerned with the relationships between science and society, during which the interaction between Russia and the West in the subjects of superfluidity and superconductivity resembled a comic opera duet of the characters at cross purposes rather than a dialogue".
70. See W. Heisenberg, 1949.
71. B. B. Goodman, 1953. The Heisenberg—Koppe theory could be interpreted in terms of an energy gap.
72. F. London, 1949a.
73. H. Fröhlich, 1961, (p. 7).
74. H. Fröhlich, 1961, (p. 7). See also H. Fröhlich, 1980.
75. H. Fröhlich, 1966, (p. 539).
76. See E. Maxwell, 1950; B. Serin, C. A. Reynolds, L. B. Nesbitt, 1950. Later W. D. Allen, R. H. Dawton, J. M. Lock, A. B. Pippard and D. Schoenberg, 1950, showed that the shape of the critical field curve for tin was independent of isotopic mass, although the magnitude of the critical field at any temperature did depend on the average isotopic mass. These experiments showed clearly that the lattice is involved in the transition. See also W. D. Allen, R. H. Dawton, M. R. H. Bar, K. Mendelssohn and J. L. Olsen, 1950.

77. H. Fröhlich, 1966, (p. 551).
78. See J. Bardeen, 1952.
79. Shafroth, following Pauli's suggestions to work further on Frohlich's theory, was the first to show (in 1951) that perturbation theory could not applied. See H. Fröhlich, 1951, and M. R. Schafroth, 1951. At the same conference held in Oxford in 1951, L. Tisza said that his ideas on superconductivity "are compatible with those of Fröhlich and Bardeen, the present treatment leads to a parameter of long range order, which the single-electron theories lead to short range order only" (p. 113).

    In an earlier paper L. Tisza, 1950, tried to develop his own quantum-mechanical model using localized "atomic" wave functions in order to construct many-electron wave functions, but he said that "It is possible that ... [Fröhlich's theory] is the definitive theory of superconductivity rendering the ideas of the present paper obsolete" (p. 725). See also J. M. Luttinger, 1950, where the London equation connecting the electric field and the supercurrent is derived on the basis of Tisza's theory. Tisza's theory was the first attempt to introduce correlations between electrons in the superconducting state systematically. It appeared later that the correlations he envisaged between electrons were too strong. By 1953, it was generally agreed that the interaction between the electrons and the lattice was responsible for superconductivity but that the wave functions proposed by Fröhlich and Bardeen were inadequate. Fröhlich himself had emphasized the need for new mathematical methods. [See H. Fröhlich, 1953, at the Lorentz—Kamerlingh Onnes Conference, where in illuminating discussions by Bohr, Heisenberg and Casimir the situation was very well summarized. (*Physica* 19, pp. 761—764)].

    Mainly to convince himself that solutions which could not be obtained from perturbations theory were possible, Fröhlich solved a one-dimensional model. This model contained several features of the present theory; metastable currents in a ring, an energy gap, etc. [See H. Fröhlich, 1954].
80. "Yet it appears that all this did not affect some of those physicists who had previously formed an opinion in a different direction. A striking example can be found in the article by W. L. Ginsburg published in 1953. ... He must have been very uncertain of the correctness of his opinion and yet unprepared to give it up, for otherwise it would be difficult to understand the angry and quite incorrect report he gives on the early papers dealing with electron—phonon interaction. Also it is not without amusement that one reads now his dogmatic prescriptions as to how work should proceed in order to produce a correct theory". H. Fröhlich, 1961, (p. 20).
81. J. Bardeen, 1963, (p. 25).
82. "I had shown then that the electron—phonon interaction whose strength can be obtained from the high temperature electric resistivity must also lead to an electron—electron interaction. The free-electron model was thus taken for granted, which implies the most of the interelectronic Coulomb energy is already considered. Later applications of this interaction have explicitly added some further Coulomb interaction between the electrons". H. Fröhlich, 1966, (p. 544).
83. Although, in the Bardeen—Pines model, the canonical transformation was chosen by perturbative methods, it was not necessary for the wave function to be chosen in this way. "This marriage of perturbation theory and a nonperturbative solution is not a happy one fundamentally, but it has proved exceedingly useful; it is a *marriage de convenance.* The effective Hamiltonian born of this was, in fact, the basis for the present microscopic theory". [G. Rickayzen, 1964, (p. 22)].
84. An early treatment of such a many-body problem without the use of perturbation theory was Bogoliubov's, of a dilute imperfect Bose gas. Bogoliubov was able to show that such a gas would have phononlike excitations and superflow properties which resembled some of the properties of liquid helium II. See Bogoliubov, 1947.

    Another many-body problem which had a bearing on the theory of superconductivity was that of the free electron gas. Bohm and Pines showed that because of the long-range character of the Coulomb force the system possesses collective density oscillations of high frequency, which are not excited in low-frequency phenomena, like superconductivity, and usually can be ignored. But, the electrons behave also as free particles with a short-range interaction between them, and, thus, the success of the independent electron model of metals is explained.

85. P. W. Anderson, 1969, p. 1349.
86. G. Rickayzen, 1965, (p. 24). Actually, the idea of bound electron pairs emerged for the first time in an attempt by Ogg, 1946, to explain the phase separation and superconductivity of metal-ammonia solutions. See also J. M. Blatt, 1964, (pp. 86—87).
87. The energy-gap model was the unifying theme of this review article, in which Bardeen suggested that one should take the complete interaction, not just the diagonal self-energy terms and use it as the basis for a theory of superconductivity.
88. J. Bardeen, 1973c, (p. 31).
89. J. Bardeen, 1963, (p. 26).
90. For a brief presentation of some of the experimental results which are explained by the theory, see D. M. Ginsberg, 1962. A detailed summary of the experimental work done before 1956 is B. Serin's, 1956.
91. J. Bardeen, 1973c, (pp. 35—36). In (pp. 36—41) of the same paper one can find interesting information on the reception of the B.C.S. theory by the scientific community. For a summary of results obtained after the proposal of the B.C.S. theory, see J. Bardeen and J. R. Schrieffer, 1961. See also J. Bardeen, 1969, and M. J. Buckingham, 1961.

Chapter 4

1. J. F. Allen, 1952, (p. 66).
2. "... it was almost a kind of sports event to liquefy all gases" [H. B. G. Casimir, 1977, (p. 171)]. "People busy with such activities were regarded much in the same way as someone who wanted to be first at the North Pole or run faster than anyone else". [F. E. Simon, 1952, (p. 1)]. See, for example, Dewar's polemic with Olszewski as to the priority on some achievements in the liquefaction of gases. [C. Olszewski, 1895; J. Dewar, 1895. See also K. Adwentowski, A. Pasternak and Z. Wojtasjek, 1957—58].
3. A very long preparation was necessary before he succeeded. By 1908, 36 articles had already appeared in his Leiden Communications "On the methods and apparatus used in the cryogenic laboratory" and "On the measurement of very low temperatures" [see R. de Bruyn Ouboter, 1986]. This methodical and systematic nature of Kamerlingh Onnes and of all his work led to the development of a laboratory (equipped with powerful liquefiers for air, hydrogen and helium and with apparatus for work in all domains of physics) rather different from the average place of research. Characteristic is the fact that, whereas Dewar liquefied hydrogen in 1898, it was not until 1906 that liquid hydrogen was produced at Leiden because "Onnes was determined not to set up a toy. His liquefier was to be a genuine machine, determined to supply a generation of physicists with as much hydrogen as they needed and, more important still, to lead to the production of helium. Thus, it came about that only two years elapsed between the completion of the hydrogen plant and the liquefaction of helium in 1908". [M. and B. Ruhemann, 1937, (p. 39)].
4. Three days later Lorentz announced Kamerlingh Onnes's achievement as follows: "le professeur Kamerling Onnes à Leiden vit une longue suite d'années d'un labeur intense, guidé par les théories de M. van der Waals, couronnée de succès par un de ces resultats qui méritent d' entre enregistrés en lettres d'or: la liquefaction de l'helium". [H. A. Lorentz, 1908, (p. 492)].
5. This is claimed by a number of authors, but there does not seem to be any evidence for it. There are no notebooks of Onnes referring to the measurements and observations of July 10, 1908. There are, however, notes for that day in the notebooks of G. Flim, the chief technician of the Laboratory. Nothing is found there concerning peculiar properties of liquid helium. Even the verbal accounts of that day do not refer to anything peculiar about liquid helium apart from its remarkable transparency. Actually liquefying helium was so important

and exciting, that everything else on that day would not have struck the group as important — since also the experimental set-up was not particularly favourable for any other observations.

6. It has to be noted here that the actual temperatures reached by pumping off the vapour are subject to a small degree of uncertainty, considering the temperature scale used by the researchers. The early results were re-calculated by Keesom on the basis of the "1932 scale" and yield the following picture. In his first liquefaction (1908) Kamerlingh Onnes reached a temperature of 1.72°K. In this and the following three attempts in 1909, 1910 and 1919 mechanical pumps were used and the temperatures attained were 1.58, 1.04 and 1.00°K respectively. Using diffusion pumps he reached in 1922 a vapour pressure of 0.013 mm Hg, corresponding to a temperature of 0.83°K, and ten years later Keesom succeeded in pumping helium down to 0.71°K. The success of the magnetic cooling method in the following year (W. F. Giauque, D. P. MacDougal, 1933), diminished interest in the attainment of very low temperatures with helium. In this kind of experiments a paramagnetic salt is cooled to as low a temperature as possible with the aid of liquid helium. A strong magnetic field is then applied, producing a rise of temperature in the substance and a consequent flow of heat to the surrounding helium, some of which is thereby evaporated. After a while, the substance is both strongly magnetized and colder. At this moment, the space surrounding the substance is evacuated. The magnetic field is now reduced to zero, and the temperature of the paramagnetic salt drops to a low value.
7. See H. Kamerlingh Onnes, 1908, (esp. p. 18), where the first liquefaction of helium is described with great precision, in full detail.
8. H. Kamerlingh Onnes, 1911a, (see p. 4).
9. H. Kamerlingh Onnes, 1913e, (p. 327).
10. H. Kamerlingh Onnes, 1913d, (p. 55).
11. H. B. G. Casimir, 1973, (p. 493).
12. H. B. G. Casimir, 1977, (p. 174).
13. Low-temperature technique is faced with three fundamental problems. First, the low temperatures must be produced, secondly, they must be measured, and thirdly, such apparatus must be devised as will allow us to bring the object to be investigated to the temperature required, to maintain it at that temperature during the experiment and to carry out the experiment under these conditions.
14. Only in 1932 was production finally brought to such a state of efficiency that the cost was $4.93 per thousand cubic feet based on a production of 1 652 000 cubic feet. [see A. Stewart, *U.S. Bureau of Mines, I.C. 6745*, 1933; see also E. Burton, H. Grayson Smith, J. Wilhelm, 1940, (pp. 9—10), on the production of helium gas]. After the First World War in 1918 only 300 liters of helium gas were left over in the Leiden laboratory. In 1919 the U.S.A. Government presented Kamerlingh Onnes 30 m^3 gas and in 1921 McLennan from Toronto came personally to Leiden with a cylinder filled with 2 m^3 of He gas for him. [See R. de Bruyn Ouboter, 1986, (p. 3)].
15. R. B. Hallock, 1982, (p. 202).
16. See E. Burton, H. Grayson Smith, J. Wilhelm, 1940, (p. 67). This classification does not represent the chronological order in which the experiments were performed.
17. H. Kamerlingh Onnes, 1922, (p. 27).
18. K. Mendelssohn, 1977, (p. 251).
19. Nearly ten years later Keesom calculated the expansion coefficient and found that "the thermodynamical consideration alluded to, lead to an interesting confirmation of the way in which Kamerlingh Onnes and Boks presented the results of their measurements on the changes in density of liquid helium in this domain". [W. H. Keesom, 1933, (p. 2)].
20. A. Th. van Urk, W. H. Keesom, H. Kamerlingh Onnes, 1925; J. I. Dana, H. Kamerlingh Onnes, 1926a; L. I. Dana, H. Kamerlingh Onnes, 1926b (abridged form in 1925).
21. L. I. Dana, H. Kamerlingh Onnes, 1926a, (p. 31, fn.1). The authors continue: "The change of density of the liquid also indicates something of the same kind". See also R. J. Donelly and A. W. Francis, 1985.

22. Kamerlingh Onnes died on 21st February 1926 and not in 1924 as M. and B. Ruhemann mention [see p. 38 of their (1937)] or in 1927 as written by Casimir in his (1973, p. 492).
23. See M. Wolfke, W. H. Keesom, 1927. In 1928 Wolfke and Keesom continued their investigations on the temperature dependance of the dielectric constant of liquid helium and they concluded that a "jump appears at a ... temperature of 2.295 on the accepted temperature scale" [M. Wolfke, W. H. Keesom, 1928, (p. 8)]. These investigations led them also to the conclusion "that the helium molecule remains the same as well in the gaseous condition, as also in liquid helium I and liquid helium II" [*ibid*, (p. 10)].
24. W. H. Keesom, M. Wolfke, 1927, (p. 22). See also W. H. Keesom 1928.
25. G. E. Mac Wood, 1938 a; G. E. MacWood, 1938b; W. H. Keesom, J. E. MacWood, 1938.
26. H. B. G. Casimir, 1973, (p. 494).
27. J. C. McLennan, H. D. Smith, J. O. Wilhelm, 1932, (p. 165).
28. W. H. Keesom, A. P. Keesom, 1932, (p. 19).
29. "So our conclusion is that the specific heat of liquid helium at about 2.19 °K falls from the value of 3.0 to a value of about 1.1 certainly within 0.002 degree, very probably even within thousandths of degree" [*ibid*., p. 25]
30. *Ibid*., (pp. 25—26). The discovery of the lambda-transition in liquid helium led Ehrenfest to consider this type of transformation in more general terms. He introduced the idea of an nth order transition when the Gibbs potential $G(p, \theta)$ has a discontinuity of the nth order partial derivatives.

    A first order transition, such as an ordinary phase transition, involves a latent heat and a change of volume, which means that, although the Gibbs free energy G is unchanged by the transition, there is a discontinuous change in each of its first order derivatives, the entropy $S = -(\partial G/\partial T)_p$ and the volume $V = (\partial G/\partial P)_T$. In a second order transition, such as the transition from helium I to helium II, the first order derivatives are continuous, but there is a discontinuous change in the second order derivatives, the specific heat at constant pressure $C_p = -T(\partial^2 G/\partial T^2)_p$, the coefficient of expansion a = $1/V\ (\partial^2 G/\partial T \partial P)$, and the compressibility $K_T = -1/V\ (\partial^2 G/\partial p^2)_T$. (P. E. Ehrenfest, 1933).

    Von Laue and others raised strong objections to Ehrenfest's point of view (see P. Epstein, 1937, section 49) but the theory found widespread application to a variety of phenomena occuring in solids. Later, rigorous statistical calculations of Onsager have revealed that the above theory is not general enough to provide a satisfactory phenomenological framework. The same criticism also applies to Landau's extension of Ehrenfest's ideas. It appears that λ-points are not discontinuities, but singularities, of the thermodynamic quantities.
31. W. H. Keesom, A. P. Keesom, 1935.
32. Again, technicians in the Leiden Lab "were no surprised. If you pass a current through a little heating spiral immersed in liquid helium then you see bubbles above 2.19 °K but you don't get bubbles at lower temperatures. But nobody had paid attention to this fact". [H. B. G. Casimir, 1973, (p. 493)].
33. W. H. Keesom, A. P. Keesom, 1936, (p. 360).
34. J. F. Allen, R. Peirls, M. Zaki Uddin, 1937.
35. They observed that the heat conductivity, if defined by the ratio (heat/temperature gradient), depends on the temperature difference and appears to become infinite with decreasing temperature difference.
36. Actually Keesom used the expression "supra heat conductivity". In Keesom and Keesom, 1936, (p. 360) we read: "Connecting this with the abrupt change of the heat conductivity in passing the lambda-point we may perhaps be justified in calling liquid helium II supra-heat-conducting".
37. C. T. Lane, 1962, (p. 33).
38. J. F. Allen, 1952, (p. 78).
39. The observations of Daunt and Mendelssohn were supported by direct interferometric measurements by E. J. Burge and L. C. Jackson (1949), who showed, moreover, that the film seemed to disappear at the λ-point. This indicated that the presence of the film could be accounted for by van der Waals forces alone.

40. K. Mendelssohn, 1972, (p. 430); see also his ref. 21 and 22.
41. F. E. Simon, 1952, (p. 1).
42. Trouton's rule, in its usual form, states that the quotient of the latent heat of evaporations per gram molecule and the absolute temperature of ebullition is about 21. This means that the change in entropy on vaporisation is about the same for all substances, a result ultimately depending on the facts that the molecular volume of all gases is the same and that the entropy of a liquid is negligible compared with that of vapour.
43. See H. Kamerlingh Onnes, 1911a and 1913e and text to footnote 9.
44. See M. Planck, 1911a.
45. W. H. Keesom 1926b and 1926c. On Thursday 1st July 1926 the editor of *Nature* received from Keesom the following telegram: "Helium solidified under a pressure of 150 atmospheres at the temperature of its boiling-point and under 28 atmospheres at 1.5° Abs. Solid helium forms transparent mass". [see *Nature*, vol. 118, July 10, 1926, p. 58].
46. F. E. Simon, 1927, (p. 808). In 1934 he pointed out that the high zero-point energy of helium was responsible for keeping the substance, under saturation pressure, in the liquid phase down to absolute zero. (see F. E. Simon, 1934).
47. See W. H. Keesom, 1927. Nernst's heat postulate (which he derived intuitively without reference to quantum theory by considering chemical equilibria and high temperatures), or the third law of thermodynamics, was originally stated as follows: For any reversible isothermal process which may take place in a condensed system *the change of entropy approaches zero as the temperature approaches the absolute zero.* (W. Nernst, 1906). In the general treatment of thermodynamic relations, based on the second law, it is possible to define only the difference of entropy between two states. Planck recognized that Nernst's postulate can be used to assign a meaning to the absolute value of the entropy; he therefore restated the third law in the form: for all states of a system (in thermal equilibrium) *the entropy approaches zero as the temperature approaches zero.* [See M. Planck, 1911b, (p. 269)].

    Through the statistical interpretation of entropy as a measure of the disorder of the system, Planck's form of the law can be given a very simple meaning: *All substances at absolute zero are in a state of perfect order.* (W. Nernst 1918; and F. E. Simon, 1927).

    Broady speaking, we may state that, at low temperature, the molecules are organized into regular periodic patterns forming a crystal. Another type of molecular order is obtained when certain atoms or molecules carrying magnetic or electric fields are oriented (ferromagnetism and ferroelectricity).
48. Indeed, a liquid with a zero entropy presented a great problem, because in a liquid the atoms are not arranged in an orderly manner and disorder means entropy. This had been shown clearly by Simon and Lange in their study of the entropy of amorphous substances. They found that amorphous glycerol has a higher entropy than crystalline glycerol and this was attributed to its disorderly structure. Simon resolved the apparent contradiction with the third law by pointing out that such substances are not in *internal thermodynamic equilibrium* and cannot therefore be treated thermodynamically. Now, here in the liquid state, there was apparently a similar configuration which had zero entropy. [See F. Simon, F. Lange, 1926].
49. The first hint of such an idea is found in W. H. Keesom and M. Wolfke, 1927, (p. 22).
50. See W. H. Keesom, 1932, (p. 51).
51. K. Clusius, paper read in Breslau, 1933, unpublished.
52. The zero-point energy may be estimated theoretically (from a consideration of the balance of forces) in two limiting cases: when the interatomic distance is less than an atomic diameter and when the interatomic distance is much greater than the atomic diameter. F. London used an interpolation formula between these limiting cases. See R. P. Dingle, 1952, (pp. 118—119).
53. F. London, 1936, (p. 580). "It is perhaps characteristic of the trend of thought at the time that F. London avoided the term 'liquid' in the title of his paper, referring to 'condensed' helium." [K. Mendelssohn, 1956, (p. 386).]
54. F. London, 1936, (p. 581).
55. *Ibid.*, (pp. 582—583).

56. F. London, 1938a, (p. 643).
57. F. London, 1939, (p. 58).
58. See H. London, 1960, (p. 39).
59. W. H. Keesom, K. W. Taconis, 1938a.
60. E. Fermi presented his quantum theory of the ideal gas in February 1926 and P. A. M. Dirac in August 1926. Both Fermi and Dirac presented essentially the same derivation for their distribution law, though Fermi worked out its consequences in greater detail. Bose—Einstein statistics were actually developed before quantum mechanics, though their full significance was not appreciated until later. In June 1924 S. Bose sent Einstein a paper in English on the derivation of Planck's radiation law, asking him to arrange for its publication in the *Zeitschrift für Physik*. Einstein translated it into German himself and added a note stating that he considered the new derivation an important contribution. The paper was published a few weeks later. Einstein immediately followed up Bose's approach by applying it to a monatomic ideal gas and in 1925 he predicted a quantum condensation of gases. See: Fermi, 1926; Dirac, 1926; Bose, 1924; Einstein, 1924 and 1925. See also S. G. Brush, 1983, (pp. 157—162).
61. H. London, 1960, (p. 39).
62. F. London's first step was analogous to Sommerfeld's treatment of the free electron gas. (See A. Sommerfeld, 1928).

    Many years before F. London's theory at a meeting of the Physical Society of London on February 14, 1930, M. C. Johnson discussed the "degeneracy" of helium gas on the basis of thermodynamic data between 4 and 6 °K, using Fermi's correction to the pressure of an ideal gas. J. E. Lennard-Jones pointed out that he should have used Bose—Einstein statistics ("the author considers only the Fermi—Dirac statistics, whereas theory indicates that helium atoms should obey the Bose—Einstein statistics. It would add to the value of his work if the author could consider the effect of the latter statistics on helium near the critical point"), but this suggestion was not followed up. [M. C. Johnson, 1929—30; see, Discussion (p. 180)].
63. A. Einstein, 1925. Uhlenbeck, in his doctoral thesis, had called Einstein's result into doubt and there the matter rested until F. London's letter to *Nature* and Uhlenbeck and Kahn's paper in *Physica* in which Uhlenbeck had withdrawn his former objection. [See G. E. Uhlenbeck, 1927; F. London, 1928a]. "If the λ-phenomenon of liquid helium had been discovered between 1925 and 1927, one would perhaps have tried at once to interpret it as the condensation predicted by Einstein" [F. London, 1939, (p. 59)].
64. F. London, 1938b, (p. 951).
65. See footnote 33.
66. F. London, 1938a, (p. 644).
67. As Sir William Bragg referred to the rather confusing situation in which the scientific community found itself after the discovery of the "peculiar" properties of liquid helium II. [see K. Mendelssohn, 1972, (p. 430)].
68. L. Tisza, 1949, (p. 2).
69. F. London, 1938b and 1939.
70. F. London, 1947, (p. 8).
71. J. F. Allen, 1952, (p. 90).
72. Kapitza's experiments were not conducted to test Tisza's model, but ostensibly to measure accurately many of the liquid helium properties.
73. P. L. Kapitza, 1941a, (p. 581).
74. "It is indeed unthinkable that Helium can move in a capillary at a speed exceeding that of a bullet". [P. L. Kapitza, 1940, (p. 24)].
75. P. L. Kapitza, 1940, (p. 24).
76. P. L. Kapitza, 1941b, (p. 638).
77. At the same time Kapitza corrected his earlier model of surface flow and substituted for it Landau's new "two-fluid" model. [L. D. Landau, 1941]. The 1941 paper is not Landau's first incursion into the problems of liquid helium; in his 1937 paper on phase transitions he had suggested that He II is a "liquid crystal" with cubic symmetry, but did not pursue this idea. [L. D. Landau, 1937a].

78. J. Frenkel, 1946, (esp. p. 308).
79. In his "quantized hydrodynamics" the macroscopic density and velocity of the fluid would be replaced by noncommuting quantum-mechanical operators. Rather than attempt to derive hydrodynamics from the Schrödinger equation, Landau was following one of the paths by which the Schrödinger equation itself could have been derived in an axiomatic treatment of quantum mechanics. "Such a theory called for an experimental test in the classic hypothetico deductive tradition even though (here as in other cases) scientists do not let theories stand or fall on the basis of experiments alone". [G. S. Brush, 1983, (p. 182)].
80. Landau assumed the ground state to be free of vortices, and attempted to prove this assumption by demonstrating that vortex motions, which would be quantized, would require the addition of a finite energy increment to the system. This increment δ was assumed to be positive; i.e., the system with vortex excitations would have an energy greater than the ground state. Although this assumption does not prove that the ground state is vortex free, it is natural to suppose that vorticity would increase the entropy of the system.
81. "This name was suggested by I. E. Tamm". [L. D. Landau, 1941, (fn. in p. 200)].
82. The statement in Keesom's book on Helium [Keesom, 1942], that in Landau's theory "phonons and rotons act the part of superfluid and normal fluid respectively" is erroneous. Actually, phonons and rotons are types of elementary excitations and, thus, they both "play the role" of normal fluid.
83. L. Landau, 1941, (p. 192). Years later he said: "I am glad to use this occasion to pay tribute to L. Tisza for introducing, as early as 1938, the conception of the macroscopical description of helium II by dividing its density into two parts and introducing, correspondingly, two velocity fields. This made it possible for him to predict two kinds of sound waves in helium II. (Tisza's detailed paper (*J. Phys. rad.* 1, 165, 350 (1940)) was not available in USSR until 1943 owing to war conditions, and I regret having missed seeing his previous short letter (*Comptes Rendus*, 207, 1035, 1186, (1938)).)". But he insisted that Tisza's "entire quantitative theory (microscopic as well as thermodynamic-hydrodynamic) is in my opinion, entirely incorrect". [L. D. Landau, 1949, (p. 474, fn.1)]. It is interesting to note here that while Landau quoted Tisza, he never referred to F. London's original work.
84. V. L. Ginzburg, 1943, (p. 305).
85. V. Peshkov, 1944; V. Peshkov, 1946.
86. F. London, 1947, (p. 13).
87. L. Tisza, 1949, (p. 2).
88. Landau was at this time unaware of Tisza's prediction of temperature waves. [See footnote 91].
89. According to Peshkov, 1946, (p. 167). "An attempt to detect the second sound by the beats in standing waves radiated by oscillating piezoquartz was undertaken in the Institute for Physical Problems by Shalnikov and Sokolov before the war, but without success".
90. S. G. Brush, 1983, (p. 184).
91. E. Lifshitz, 1944, (p. 241).
92. V. Peshkov, 1944. It is interesting to note that in 1940 Ganz sent a heat pulse down a long capillary of He II and estimated its velocity to be of the order of 100 m/sec. "Although it was not then recognized as such, this must be considered to be the first observation of a travelling temperature wave in helium II". [J. F. Allen, 1952, (p. 90)].
93. In his 1944 paper Peshkov reported that the velocity of second sound was 19 ± 1 m/s at 1.4 °K, compared with 26 m/s estimated by Lifshitz; in his 1946 experiment he found that the velocity reaches maximum of 20.3 m/s at about 1.65 °K and then slowly falls. Tisza and London, on the basis of the theory of the Bose—Einstein condensation, predicted that this fall should continue below 1 °K.
94. V. Peshkov, 1946, (p. 185). At this point it would seem that the second-sound data favoured Tisza's theory over Landau's, but as Peshkov said "Only after experiments at lower temperatures will be carried out will it be possible to determine whether . . . the microscopic theory of helium II should be modified in some manner" [*ibid*, p. 186]. Landau himself stated that "The experimental data which are available at present are yet insufficient to disprove Tisza's assertion, because of the comparatively small role of the phonons in the

temperature region explored. But I have no doubt whatever that at temperatures 1.0—1.1 °K the second sound velocity will have a minimum and will increase with the further decrease in temperature. This follows from the values of the thermodynamic quantities of helium II calculated by me". [L. Landau, 1949, p. 476]. Tisza accepted the challenge and maintained that his own formula for second-sound velocity should be more accurate at lower temperatures. [See L. Tisza. 1949]. Peshkov, by extending the range of the experiments to 1.03 °K, was able to show a slight increase in the velocity at the lowest temperature. The question was, however, definitely settled by Pellam and Scott who studied second sound pulses in magnetically cooled helium. They found that by cooling they could raise the velocity of second sound to 34 m/s. This spectacular increase left little doubt, that, as regards to the propagation of second sound, the prediction of Landau appeared to be the more probable one. (V. Peshkov, 1948,; J. P. Pellam, R. B. Scott, 1949). However, for finite temperatures there was a discrepancy between theory and experiment.

95. Although this discrepancy is not very large, it is too large to be attributed to the inaccuracy of the experimental data on the thermodynamic quantities of helium II". [L. Landau, 1947, (p. 243)].
96. *Ibid.*, (p. 245).
97. *Idem.*
98. N. Bogoliubov, 1947, (p. 247).
99. L. Tisza, 1949, (pp. 2—3).
100. J. F. Allen, 1952, (p. 92).
101. F. London, 1949b, (p. 696).
102. *Idem.* The enormous success of nuclear physics in the 40s has made it possible to obtain He^3 in quantities sufficient for experimentation. In 1948 at Los Alamos it was shown that He^3 liquefies at 3.2 °K, and a new quantum liquid was made available to physicists. (S. Sydoriak, E. Grilly, E. Hammel, 1949). The fact that He^3 could be obtained as a liquid at atmospheric pressure might itself be considered a point against the London—Tisza view, since London himself stated that owing to its high zero point energy it is "almost certain that pure He^3 cannot exist in a liquid phase at any temperature . . ." and this view was shared by Tisza [see F. London, O. Rice, 1948, (p. 1193); L. Tisza, 1948, (p. 26)].

    In 1956, Landau developed a separate theory for Fermi liquids (L. Landau, 1956 and 1957). When superfluidity was finally discovered in He^3 in late 1971, Landau's theory was found best suited to describe it. [See O. Osheroff, R. Richardson, D. Lee, 1972; T. Alvesalo, Y. Anufriyev, H. Collan, O. Lounasmaa, P. Wennerstrom, 1973]. For a description of the superfluid phases of He^3 including both experimental observations and the essential ideas behind the theory see N. D. Mermin, D. M. Lee, 1976].
103. On the role of intermediate concepts in the development of theories of liquid helium II and many-particle physics, see F. Reif, 1968; K. Gavroglu, Y. Goudaroulis, 1986; Y. Goudaroulis, 1988.
104. N. Bogoliubov, 1947, (p. 247).
105. *Ibid.*, p. 248. It should be emphasized that a quasiparticle is a purely theoretical construct, having nothing to do not with an individual helium atom, but with motion of the liquid as a whole. Nevertheless, the behaviour of a quasi-particle gas is remarkably like that of a real gas, with two important differences. The first of these concerns the relation between the energy and momentum of a quasi- particle. This relation reflects the properties of the modes of motions of the whole liquid and it quite unlike the corresponding relation for a real particle. The second difference concerns the number of particles in a given sample of material. In a real gas the number is fixed. In helium II it depends on the temperature. At absolute zero there are no quasi-particles, and their number increases as the temperature, and hence energy of the fluid, is raised.
106. S. G. Brush, 1983, (p. 189).
107. N. Bogoliubov, 1947, (p. 248). At that time a theory of liquid helium not based on analogies with gas models but considered as a limiting case of the theory of liquids had been given by Green. His theory was based on the quantal formulation of the kinetic theory of liquids of

Born and Green. In such a kinetic theory of a quantum liquid the states of angular momentum $l = 2, 4, \ldots$ give no contribution at very low temperatures. The radial distribution function then obtained is very different from that of a normal liquid. It is supposed that the transition is located at that temperature for which the last non-vanishing state of angular momentum begins to disappear. In a sense this explanation of the existence of He II is similar to that suggested by Landau. According to him the curious properties of He II are to be correlated with the insufficiency of quantized vortex (rotational) states, whilst according to Green they are to be correlated with the insufficiency of quantized states with non-zero relative angular momenta. It is a feature of the theory that the classical conceptions of temperature and pressure are no longer applicable. This first law of thermodynamics takes different forms depending on whether the liquid is in steady motion or undergoing periodic displacements. The later case leads to the possibility of the excitation of thermal waves which transfer heat energy and liquid bulk in opposite directions; to these are ascribed the transport phenomena of He II.

Another attempt to discuss the problem of liquid helium based on a theory of liquids was Prigogine and Philippot's generalization of the cell model of an ordinary liquid developed by Lennard-Jones and Devonshire. In a series of papers they have a model of the thermal properties of helium II, but no attempt was made to discuss its dynamical properties. The model shows the analogue of a Bose—Einstein condensation.

It shows roughly how a λ-type of specific heat anomaly may arise in He^4, but not be found in He^3, and leads to a negative expansion coefficient below the λ- point.

108. He cites in evidence Taconis's assumption that He^3 dissolves only in the normal part of He II. Taconis made this suggestion in 1949 in an attempt to explain his experimental result that the effect on the vapour pressure of He II caused by the addition of He^3 is much greater than that predicted by Raoult's law.

109. Daunt and Mendelssohn have drawn attention to the analogous behaviour of He II and superconductors, and extended the two-fluid conception to both phenomena. They suggested that both are due to a new state of aggregation, composed of "z-particles", in which frictionless transport is closely associated with zero entropy without order in coordinate space. [see K. Mendelssohn, 1945]. In 1946, Macleod and Yeabsley have put forward the theory that He II is a mixture of ordinary He I and the intermediate form of a fluid indicated by van der Waals' equation, and Benedicks has suggested that the properties of He II may be explained in terms of an allotropic transformation at the lambda-point arising from the ionization of the helium atoms by frictional electricity at the walls of the vessel.

110. F. London, 1951, (pp. 2—3).

111. R. Feynman, 1955, (p. 25).

112. R. Feynman, 1953a, (p. 1116). See also R. Feynman, 1953b.

113. R. Feynman, 1953c.

114. R. Feynman, 1954.

115. R. Feynman, M. Cohen, 1956.

116. L. Onsager, Remark at a Low Temperature Physics Conference at Shelter Island in 1948, Published in *Nuovo Cimento*, supplement, 6, 1949, (p. 249).

117. R. Feynman, 1955, (p. 45).

118. It has to be noted here that the consideration following from the Landau spectrum gave a critical velocity $v_c = 8000—24\,000$ cm/s. An experimental evidence of the existence of vortex lines was given first by Hall and Vinen in 1956, and in 1958 Vinen at the Royal Society Mond Laboratory in Cambridge observed the quantized vortex lines predicted by Onsager and Feynman [see W. Vinen, 1958; 1961. See also G. Rayfield, F. Reif, 1963].

Chapter 5

1. We will not repeat in this chapter the references concerning the properties of superconductivity and superfluidity with which we dealt in chapters 3 and 4.

2. H. Kamerlingh Onnes, 1913a, p. 13.
3. W. J. De Haas, J. Voogd, 1931a, p. 21.
4. J. Bardeen, 1963, p. 21.
5. R. B. Dingle, 1952, p. 147.
6. F. London, 1961, pp. 2—3.
7. F. and H. London, 1935, p. 74.
8. *Ibid.*, p. 73.
9. E. F. Burton, 1940, p. 331.
10. F. London, 1961, p. 4.
11. E. F. Burton, H. Grayson Smith, J. O. Wilhelm, 1940, p. 336.
12. F. London, 1947, p. 3.
13. L. D. Landau, 1941, pp. 209—210.
14. F. London, 1947, p. 13.
15. *Ibid.*, p. 13.
16. L. D. Landau, 1949, p. 474.
17. R. P. Feynman, M. Cohen, 1956, p. 1189.
18. K. Mendelssohn, 1977, p. 114.

Bibliography

(In the Bibliography *CPL* stands for *Communications from the Physical Laboratory at the University of Leiden*)

Adwentowski, K., Pasternak, A., Wojtasjek, Z., 1957—58, "Who of the two: Dewar or Olszewski?" *Kwartalnik Historii Nauki i Techniki*, special issue, 77—95.

Allen, J. F., 1952, "Liquid helium", in F. Simon *et al.*, 1952, 66—94.

——, Jones, H., 1938, "New phenomena connected with heat flow in helium II", *Nature*, **141**, 243—244.

——, Misener, A. D., 1938, "Flow of liquid helium II", *Nature*, **141**, 75.

——, Peirls, R., Zaki Uddin, M., 1937, "Heat conduction in liquid helium", *Nature*, **140**, 62—63.

Allen, W. D., Dawton, R. H., Lock, J. M., Pippard, A. B., and Schoenberg, D., 1950, "Superconductivity of tin isotopes", *Nature*, **166**, 1071.

——, Dawton, R. H., Bar, M. R. H., Mendelssohn, K. and Olsen J. L., 1950, "Superconductivity of tin isotopes", *Nature*, **166**, 1071—1072.

Alvesalo, T., Anufriyev, Y., Collan, H., Lounasmaa, O., Wennnerstrom, P., 1973, "Evidence for superfluidity in the newly formed phases of ^{3}He", *Physical Review Letters*, **30**, 962.

Amagat, M. E. H., 1896 a, "Verification d'ensemble de la loi des etats correspondants de Van der Walls", *Comptes rendus des seances de l'Academie des Sciences*, **CXXIII**, 1—6.

——, 1896b, "Sur le loi des etats correspondants de Van der Waals et la determination des constantes critique", *Comptes rendus des seances de l'Academie des Sciences*, **CXXIII**, 7—12.

Anderson, P. W., 1969, "Superconductivity in the past and the future" in R. D. Parks, 1969, Vol. 2, 1343—1358.

Andronikashvilli, E., 1946, "A direct observation of two kinds of motion in helium II", *Journal of Physics* (USSR), **10**, 201—206.

Atkins, K. R., *1952*, "Wave propagation and flow in liquid helium II", *Philosophical Magazine Supplement*, **1**, 169—208.

Baltas, A., 1986, "Ideological 'assumptions' in physics: social determinations of internal structures", *PSA 1986*, Vol. 2, to be published.

——, Gavroglu, K., 1980, "Modification of Popper's tetradic schema and the special theory of relativity", *Zeitschrift fur allgemeine Wissenschaftstheorie*, **XI**, 213—237.

Bantz, D. A., 1980, "The structure of discovery: Evolution of structural accounts of chemical bonding", in Nickles T., 1980b, 291—329.

Bardeen, J., 1940, "Electrical conductivity of metals", *Journal of Applied Physics*, **11**, 88—111.

——, 1950a, "Zero Point Vibrations and Superconductivity", *Physical Review*, **79**, 167—168.

——, 1950b, "Wave Functions for Superconducting Electrons" *Physical Review*, **80**, 567—574.

——, 1952, "Superconductivity and lattice vibrations", in *Low-Temperature Physics*, Proceedings of the NBS Semicentennial Symbolisium on Low-Temperature Physics, March 27—29, 1951, National Bureau of Standards Circular 519.

——, 1956, "On superconductivity", in S. Flugge, ed., *Encyclopedia of Physics*, Vol. XV.
——, 1958, "Theory of superconductivity", (Kamerlingh Onnes Conference, Leiden), *Physica*, **XXIV**, 27—34.
——, 1963, "Developments of concepts in superconductivity", *Physics Today*, **16**(1), 19—28.
——, 1969, "Advances in superconductivity", *Physics Today*, **22**(10), 40—46.
——, 1973a, "Electron-phonon interactions and superconductivity", *Physics Today*, **26**(7), 41—47.
——, 1973b, "Electron-phonon interactions and superconductivity", in H. Haken, M. Wagner eds., *Cooperative Phenomena*, Berlin: Springer-Verlag, 63—93.
——, 1973c, "History of superconductivity research" in B. Kursunoglu, A. Perlmutter (eds.), *Impact of Basic Research on Technology*, New York: Plenum Press, 15—57.
——, Cooper, L. N., Schrieffer, J. R., 1957a, "Microscopic theory of superconductivity", *Physical Review*, **106**, 162—164.
——, 1957b, "Theory of superconductivity", *Physical Review*, **108**, 1175—1200.
Becker, R., Heller, G., Sauter, F., 1933, "Uber die Stromverteilung in einer supraleitenden Kugel", *Zeitschrift fur Physik*, **85**, 772—787.
Becquerel, J., 1912, *Comptes rendus hebdomadaires des Séances de l'Academie des Sciences de Paris*, **154**, 1795.
Benedicks, C., 1916, "Beiträge zur Kenntnis der Elektrizitätsleitung in Metallen und Legierungen", *Jahrb.d. Rad.u.Elekt.*, **13**, 351—395.
Bennewitz, K., Simon, F., 1923, "Zur Frage der Nullpunktes-energie", *Zeitschrift fur Physik*, **16**, 183—199.
Blatt, J. M., 1964, *Theory of Superconductivity*, New York: Academic Press.
Bloch, F., 1928, "Uber die Quanten-mechanic der Elektronen in Kristallgittern", *Zeitschrift fur Physik*, **52**, 555—600.
——, 1980, "Memories of electrons in crystals", *Proceedings of the Royal Society of London*, **371A**, 24—27.
Bogoliubov, N., 1947, "On the theory of superfluidity", *Journal of Physics* (USSR), **11**, 23—32. Reprinted in Galasiewicz, 1971, 247—267.
Borelius, G., 1935, "Physikalische Eigenschaften der Metalle" in *Handbuch der Mettalphysik* edited by G. Masing, Leipzig, vol. 1.
Bose, S. N., 1924, "Plancks Gesetz und Lichtquantenhypothese", *Zeitschrift fur Physik*, **26**, 178—181.
Breit, G., Kamerlingh Onnes, H., 1923, "Magnetic researches XXVI. Measurements of magnetic permeabilities of chromium cloride and gadolinium sulphate at the boiling point in liquid hydrogen in alternating fields of frequency 369 000 per second", *CPL*, No 168c, 21—31.
Bridgman, M. P. W., 1917a, "The resistance of metals under pressure", *Proceedings of the National Academy of Science*, **3**, 10—12.
——, 1917b, "Theoretical considerations on the nature of metallic resistance with especial regard to the pressure effect", *Physical Review*, **9**, 269—289.
——, 1921, "The discontinuity of resistance preceding supraconductivity", *Washington Academy of Sciences Journal*, **11**, 455—459.
——, 1924, "Rapport sur les phenomenes de conductibilite dans les metaux et leur explication theorique", *Conductibilité électrique des metaux et problème connexés*, rapports et discussion du 4ème Conseil Solvay, Avril 1924, (Paris, 1927), 67—114.
Brillouin, L., 1935, "Supraconductivity and the difficulties of its interpretation", *Proceedings of the Royal Society of London*, **A152**, 19—21.
Brush, S. G., 1983, *Statistical physics and the atomic theory of matter, from Boyle and Newton to Landau and Onsager*, Princeton: Princeton University Press.
Buckingham, M. J., 1961, "Epilogue: Theoretical developments 1950—1960", in F. London, 1961, 156—168.
Burge, E. J., Jackson, L. C., 1949, "Thickness of the helium II film", *Nature*, **164**, 660—661.
Burton, E. F., 1934, *Superconductivity*, Toronto: Toronto University Press.
——, Grayson Smith, H., Wilhelm, J. O., 1940, "*Phenomena at the temperature of liquid helium*, New York: Reinhold.

Cartwright, N., 1983, *How the Laws of Physics Lie*, Oxford: Clarendon Press.
Casimir, H. B. G., 1973, "Superconductivity and superfluidity" in J. Mehra, 1973, 481—498.
——, 1977, "Superconductivity", in *History of Twentieth Century Physics*, Ch. Weiner, 1977, 170—181.
——, 1983, *Haphazard Reality*, New York: Harper and Row.
Chandrasekhar, B. S., 1969, "Early experiments and phenomenological theories", in R. D. Parks, 1969, Vol. I, 1—49.
Christidis, Th., Goudaroulis, Y., Mikou, M., 1987, "The heuristic role of mathematics in the initial developments of superconductivity theory", *Archive for History of Exact Sciences*, **37**, 183—191.
Clay, J., 1908, "On the change with temperature of the electrical resistance of alloys at very low temperatures", *CPL*, 107d, 31—51.
Clerke, A. M., 1901, "Low-temperature research at the Royal Institution", *Royal Institution, Proceedings*, **16**, 699—718.
Cohen, E., 1927, "Kamerlingh Onnes memorial lecture", *Chemical Society Journal*, 1193—1209.
Cohen, R. S., Feyerabend, P., Wartofsky, M., eds., 1976, *Essays in Memory of Imre Lakatos*, Dordrecht: Reidel.
Cooper, L., 1973, "Microscopic quantum interference in the theory of Superconductivity", (Nobel Lecture), *Science*, **181**, 909—916.
Crommelin, C. A., 1910, "Isotherms of monoatomic substances and their binary mixtures IV. Remarks on the preparation of argon V. Vapour pressures above −140 °C, critical temperature and critical pressure of argon", *CPL*, No 115, 3—16.
——, 1913, "Isothermals of monoatomic substances and their binary mixtures. XV. The vapour pressure of solid and liquid argon from the critical point down to −206 °C", *CPL*, 138c, 21—32.
——, 1914a, "Isothermals of monoatomic substances and their binary mixtures XVI. New determination of the vapour-pressures of solid argon down to −205 °C, *CPL*, No 140a, 3—6.
——, 1914b, "Les recentes travaux du laboratoire cryogène de Leiden", *La génie civile*, tome *LXIV*, No. 13, 245—250.
——, 1928, "Rapport sur l'ensemble des recherches de feu M. le professeur H. Kamerlingh Onnes, aux températures de l'hélium liquide", (Cinquième Congrès International du Froid, Rome), *CPL* (suppl.), 63b, 17—40.
——, Gibson, R. O., 1927, "The vapour pressures of solid and liquid neon", *CPL*, No 185b, 15—20.
——, Martinez, J. P., Kamerlingh Onnes, H., 1919, "Isothermals of monoatomic substances and their binary mixtures XX. Isothermals of neon from +20°C to −217°C", *CPL*, 154a, 1—13.
Crowther, J. G., 1934, "Near absolute zero", *Scientific American*, **151**(12), 300—302.
Curie, M. P., 1895, "Propriétés magnétiques des corps a diverses temperatures", *Annales de Chimie et de Physique*, **5**(7), 289—405.
Dahl, P. F., 1984, "Kamerlingh Onnes and the discovery of superconductivity: The Leyden years, 1911—1914", *Historical Studies in the Physical Sciences*, **15**(1), 1—37.
——, 1986, "Superconductivity after World War I and circumstances surrounding the discovery of a state **B** = 0", *Historical Studies in the Physical Sciences*, **16**(1), 1—58.
Dana, L. I., Kamerlingh Onnes, H., 1926a, "Further experiments with liquid helium. BA. Preliminary determinations of the latent heat of vaporization of liquid helium", *CPL*, 179c, 23—34.
——, 1926b, "Further experiments with liquid helium. BB. Preliminary determinations of the specific heat of liquid helium", *CPL*, 179d, 37—45.
Daunt, J. G., Mendelssohn, K., 1938, "Transfer of Helium II on Glass", *Nature*, **141**, 911—912.
——, 1939a, "Surface transfer in Liquid Helium II", *Nature*, **143**, 719—720.
——, 1939b, "The transfer effect in liquid He II. 1. The transfer phenomena", *Proceedings of the Royal Society of London*, **A150**, 423—439.
——, 1939c, "The transfer effect in liquid He II. 2. Properties of the transfer film", *Proceedings of the Royal Society of London*, **A170**, 439—450.
——, Smith, R. S., 1954, "The problem of liquid helium some recent aspects", *Reviews of Modern Physics*, **26**, 172—236.

De Bruyn Ouboter, R., 1986, "Superconductivity: Discoveries during the early years of low temperature research at Leiden 1908—1914", Paper presented at the *H. Kamerlingh Onnes Symposium on the Origins of Applied Superconductivity*, October 1, Maryland, U.S.A.

De Boer, J., 1974 "Van der Waals in his time and the present revival", *Physica*, **73**, 1—27.

De Haas, W. J., 1911, "Isotherms of diatomic substances and of their binary mixtures. VIII. Control measurements with the volumenometer", *CPL*, No 121a, 1—18.

——, 1912, "Isotherms of diatomic gases and of their binary mixtures. X. Control measurements with volumenometer of the compressibility of hydrogen at 20 K", *CPL*, No 127a, 1—8.

——, 1933, "Supraleiter in Magnetfeld", *Leipziger Vortrage* 59—73.

——, Bremmer, H., 1931, "Conduction of heat of lead and tin at low temperatures", *CPL*, 214d.

——, 1932, "Thermal conductivity of indium at low temperatures", *CPL*, 220b, 5—11.

——, Voogd, J., 1931a, "On the steepness of the transition curve of supraconductors", *CPL*, 214c.

——, 1931b, "The magnetic disturbance of the superconductivity of single crystal wires of tin", *CPL*, 212c, 29—36.

——, Jonker, J. M., 1934, "Quantitative Undersuchung ubereinen moglichen Einfluss der Achsenorientierung auf die magnetische Ubergangsfigur", *CPL*, 229c.

——, Van Alphen, P. M., 1933, "Change of the resistance of metals in a magnetic field at low temperatures", *CPL*, 225a, 1—15.

De Haas-Lorentz, G. L., 1925, "Iets over het mechanisme van inductieverschijnselen", *Physica*, **5**, 384—388.

Dewar, J., 1895, "On the liquefaction of gases", *Philosophical Magazine*, **39**, 298—305.

——, 1901, "The nadir of temperature and allied problems", (Bakerian lecture, Royal Society, June 13, 1901), *Proceedings of the Royal Society of London*, **LXVIII**, 360—366.

——, Fleming, J. A., 1893, "The electrical resistance of metals and alloys at temperatures approaching the absolute zero", *Philosophical magazine*, **XXXVI**, 271—299.

——, 1896, "On the electrical resistivity of pure mercury at the temperature of liquid air", *Proceedings of the Royal Society*, **LX**, 76—81.

Dicke, R. H., 1961, "Mach's principle and equivalence", *Proceedings of the International School of Physics "Enrico Fermi" course 20*, 1—50, New York: Academic Press.

Dingle, R. B., 1952, "Theories of helium II", *Advances in Physics*, **1**, No 2, 111—168.

Dirac, P. A. M., 1926, "On the theory of quantum mechanics", *Proceedings of the Royal Society of London*, **A112**, 661—677.

D[onnan], F. G., 1926/7, "Heike Kamerlingh Onnes", *Proceedings of the Royal Society of London*, **113A**, i—iv.

Donnelly, R. J., Francis, A. W., eds., 1985, *Cryogenic Science and Technology, Contributions by Leo I. Dana*, with a commentary by the editors, Union Carbide Corporation.

Dorfman, J., 1923, "Einige Bemerkungen zur Kenntnis des Mechanismus magnetischer Erscheinungen" *Zeitschrift für Physic*, **17**, 98—109.

——, 1933, "Bemerkungen zur Theorie der Supraleitfähigkeit", *Phys. Zeit. der Sowjetunion*, **3**, 366—380.

——, Jaanus, R., 1929 "Die Rolle der Leitungselektronen beim Ferromagnetismus I" *Zeitschrift für Physic*, **54**, 277—288.

——, Kikoin I., "Die Rolle der Leitungselektronen beim Ferromagnetismus II", *Zeitschrift für Physik*, **54**, 288—296.

Douglass, D., Schmidt, R. W., eds., 1964, "Proceedings of the International Conference on the Science of Superconductivity" *Reviews of Modern Physics*, **36**, 1—331.

Drude, P., 1900, "Zur Elektronentheorie der Metalle", *Annalen der Physik*, **1**, 566—613.

Ehrenfest, P., 1933, "Phasenumwandlungen in uelichen und erweitert Sinn, Glassifiziert nach den entsprechenden Singula ritaten des thermodynamischen Potentials", *CPL*. (*suppl.*) 75b.

Einstein, A., 1907, "Plancksche Theorie des Strahlung und die Theorie der spezifischen Warme", *Annalen der Physik*, ser. 4, **17**, 549—560.

——, 1924, "Quantentheorie des einatomigen idealen Gases", *Situngsberichte der Preussischen Akademie der Wissenschaften, Physikalisch-Mathematische Klasse*, 261—267.

——, 1925, "Quantentheorie des einatomigen idealen Gases", *Sitzungsberichte der Preussischen Akademie der Wissenschaften, Physikalisch-Mathematische Klasse*, 3—16.

Epstein, P., 1937, *Textbook of Thermodynamics*, New York: Willey.

Feigl, H., Maxwell, G., eds., 1958, *Current Issues in the Philosophy of Science*, New York: Holt, Reinhart, Winston.

Fermi, E., 1926, "Uber die Wahrscheinlichkeit der Quantenzustande" *Zeitschrift fur Physik*, **26**, 54—56.

Feyerabend, P., 1975, *Against Method*, London: New Left Books.

Feynman, R., 1953a, "The λ-transition in liquid helium", *Physical Review*, ser. 2, **90**, 1116—1117.

——, 1953b, "Atomic theory of the λ-transition in helium", *Physical Review*, ser. 2, **91**, 1291—1301.

——, 1953c, "Atomic theory of liquid helium near absolute zero", *Physical Review*, ser. 2, **91**, 1301—1308.

——, 1954, "Atomic theory of the two-fluid model of liquid helium", *Physical Review*, ser. 2, **94**, 262—270.

——, 1955, "Application of quantun mechanics to liquid helium", in C. J. Gorter, ed., I, 17—53.

——, Cohen, M., 1956, "Energy spectrum of the excitations in liquid helium", *Physical Review*, ser. 2, **102**, 1189—1204.

Fowler, R. H., Jones, H., 1938, "The Properties of a perfect Bose—Einstein gas at low temperatures", *Proceedings of the Cambridge Philosophical Society*, **34**, 573—576.

Franklin, A., 1986, *The Neglect of Experiment*, Cambridge: Cambridge University Press.

Frenkel, J., 1946, *The Kinetic Theory of Liquids*, Oxford: Oxford University Press.

Fröhlich, H., 1937, "Zur Theorie des λ-Punktes des Heliums", *Physica*, **4**, 639—644.

——, 1950a, "Theory of superconducting state, I. The ground state at the absolute zero of temperature", *Physical Review*, **79**, 845—856.

——, 1950b, "Isotope effect in superconductivity", *Proceedings of the Physical Society (London)*, **A63**, 778.

——, 1951, "Theory of the superconductive state", R. Powers, ed., *Proceedings of the International Conference on Low Temperature Physics*, held in Oxford, August 22—28, 110—112.

——, 1953, "Superconductivity and lattice vibrations", (Lorentz—Kamerlingh Onnes Conference, Leiden, June 1953), *Physical*, **XIX**, No 9, 755—761; Discussion, 761—764.

——, 1954, "On the theory of superconductivity: The one dimensional case", *Proceedings of the Royal Society (London)*, **A223**, 296—305.

——, 1961, "The theory of superconductive state", *Reports on Progress in Physics*, **24**, 1—23.

——, 1966, "Superconductivity and the many body problem", in R. E. Marshak, 1966, 539—552.

——, 1980, "Recollections of the development of solid state physics", *Proceedings of the Royal Society (London)*, **A371**, 102—103.

——, Pelzer, H., Zienau, S., 1950, "Properties of slow electrons in polar materials", *Philosophical Magazine*, **41**, 221—242.

Galasiewicz, Z. M., 1970, *Superconductivity and Quantum Fluids*, Oxford: Pergamon Press.

——, 1971, *Helium 4*, Oxford: Pergamon Press.

Gallison, P., 1987, *How Experiments End*, Chicago: Chicago University Press.

Gavroglu, K., 1976, "Research guiding principles in modern physics: Case studies in elementary particle physics", *Zeitschrift für allgemeine Wissenschaftstheorie*, **VII**, 233—248.

——, 1985, "Popper's tetradic schema, progressive research programs and the case of parity violation in elementary particle physics 1953—1958", *Zeitschrift für allgemeine Wissenschaftstheorie*, **XVI**, 261—285.

——, 1986, "Theoretical frameworks for theories of gravitation: A case of a 'sui generis' research program", *Methodology of Science*, **19**, 91—124.

——, 1989, "Observability and Simplicity: When are particles elementary?" (to appear in *Synthese*).

——, Goudaroulis, Y., 1984, "Some methodological and historical considerations in low temperature physics: the case of superconductivity 1911—1957", *Annals of Science*, **41**, 135—149.

——, 1985, "From the history of low temperature physics: Prejudicial attitudes that hindered the initial development of superconductivity theory", *Archive for History of Exact Sciences*, **32**, 377—383.

——, 1986, "Some methodological and historical considerations in low temperature physics II:

The case of superfluidity", *Annals of Science*, **43**, 137—146.

——, 1988, "Heike Kamerlingh Onnes' researches at Leiden and their methodological implications", to be published in *Studies in History and Philosophy of Science*.

——, Nicolacopoulos, P., eds., 1988, *Proceedings of Conference "Criticism and the Growth of Knowledge: 20 Years after"* (Thessaloniki 1986), to be published by Reidel.

Giauque, W. F., MacDougal, D. P., 1933, "Attainment of Temperature Below 10 Absolute by Demagnetization" *Physical Review*, **43**, 768.

Ginsberg, D. M., 1962, "Experimental foundations of the BCS theory of superconductivity", *American Journal of Physics*, **30**, 433—438.

——, 1964, "Resource letter Scy-1 on superconductivity", *American Journal of Physics*, **32**, 85—87.

——, 1970, "Resource letter Scy-2 on superconductivity", *American Journal of Physics*, **38**, 949—955.

Ginzburg, V. L., 1943, "Scattering of light in helium II", *Journal of Physics* (USSR), **7**, 305—306. Reprinted in Z. Galasiewicz, 1971.

——, 1955, "On the theory of superconductivity", *Nuovo Cimento*, ser. 10, **2**, 1234—1250.

Gooding, D., Pinch, T., Schaffer, S., eds., 1988, *The Uses of Experiments*, Cambridge: Cambridge University Press.

Goodman, B. B., 1953, "The thermal conductivity of superconducting tin below 1 °K", *Proceedings of the Physical Society* (*London*), **A66**, 217—222.

Gorter, C. J., 1933a, "Some remarks on the thermodynamics of supraconductivity", *Archives du Musee Teyler*, ser. III, Vol. VII, 378—387.

——, 1933b, "Theory of supraconductivity", *Nature*, Dec. 16, 931.

——, 1949, "The two-fluid model of liquid helium II", *Proceedings of the International Conference on Low Temperature Physics, at MIT*, 4—6.

——, 1953, "Some facts about superconductivity", (Lorentz—Kamerlingh Onnes Conference, Leiden, June 1953), *Physica*, **XIX**, No 9, 745—752; Discussion, 752—754.

——, 1964, "Superconductivity until 1940 in Leiden and as seen from there", *Reviews of Modern Physics*, **36**, 3—7.

——, 1967, "Bad luck in attempts to make scientific discoveries", *Physics Today*, **20**(1), 76—81.

——, Casimir, H., 1934a, "Zur Thermodynamik des supraleitenden Zustandes", *Physikalische Zeitschrift*, **35**, 963—966.

——, 1934b, "On supraconductivity I", *Physica*, **I**, 306—320.

Goudaroulis, Y., 1988, "Many-particle physics: Calculational complications that become a blessing for methodology", in Gavroglu, K., Goudaroulis, Y., Nicolacopoulos, P., forthcoming.

Gross, D., 1986, "Heterotic string theory" in G. Lazarides, Q. I. Shafi, 1986, 1—20.

Guillien, R., 1936, "La laboratoire cryogene de Leyden", *Revue generale des sciences pures et appliquees*, **XLVII**, No 4, 100—106.

Hacking, I., 1983, *Representing and Intervening*, Cambridge: Cambridge University Press.

Hallock, R. B., 1982, Resource Letter SH-1: Superfluid Helium, *American Journal of Physics*, **50**, 202—212.

Hanson, N. R., 1958a, *Patterns of Discovery*, Cambridge: Cambridge University Press.

——, 1958b, "Is there a logic of scientific discovery?", in H. Feigl, G. Maxwell, 1958, 20—42.

Heisenberg, W., 1949, "The electron theory of superconductivity", lecture delivered at Cambridge in 1948, published in W. Heisenberg, *Two Lectures*, Cambridge: Cambridge University Press.

——, 1973, "Development of concepts in the history of quantum theory" in Mehra, J., 1973, 264—275.

Hempel, C. G., 1966, *Philosophy of Natural Science*, Englewood Cliffs, New Jersey: Prentice Hall.

Hoddeson, L., Baym, G., Eckert, M., 1987, "The Development of the Quantum Mechanical Theory of Metals", *Reviews of Modern Physics*, **59**, 287—328.

Holton, G., 1973, *Thematic Origins of Scientific Thought: Kepler to Einstein*, Cambridge Mass: Harvard University Press.

Jackson, L. C., 1939, "Superconductivity", *Reports on Progress in Physics*, **VI**, 335—344.

——, 1962, "*Low Temperature Physics*", 5th edition revised, London: Methuen.

Johnson, M. C., 1929—30, "A method of calculating the numerical equation of state for helium below 6° absolute, and of estimating the relative importance of gas degeneracy and interatomic forces", *Proceedings of the Physical Society of London*, **XLII**, 170—179; Discussion 179—180.

Jones, H., 1939, "The Properties of liquid helium", *Reports of Progress in Physics*, **VI**, 280—296.

Kaischew, R., Simon, F., 1934, "Some thermal properties of condensed helium", *Nature*, **133**, 460.

Kamerlingh Onnes, H., 1879, *Nieuwe bewijzen voor de aswenteling der aarde* (New proofs of the rotation of the earth), Thesis, University of Groningen.

——, 1882, "Théorie générale de l'état fluide", *Archives Neerlandaises*, **XXX**, 1—36.

——, 1984, "On the cryogenic laboratory at Leiden and on the production of very low temperatures", *CPL*, No 14, 1—30.

——, 1896, "Remarks on the liquefaction of hydrogen, on thermodynamical similarity, and on the use of vacuum vessels", *CPL*, No 23, 1—23.

——, 1901a, "Expression of the equation of state of gases and liquids by means of series", *CPL*, No 71, 1—25.

——, 1901b, "Ueber die Riehenentamickelung fur die Zustands-gleichung der Gase und Flussigkeiten", *CPL*, No 74, 1—15.

——, 1904, "The importance of accurate measurements at very low temperatures", *CPL* (*suppl.*), No 9.

——, 1908, "The liquefaction of helium", *CPL*, 108, 3—23.

——, 1909, "Isotherms of monoatomic gases and their binary mixtures. III. Data concerning neon and helium", *CPL*, No 112.

——, 1911a, "Further experiments with liquid helium A. Isotherms of monoatomic gases etc. VIII. Thermal properties of helium B. On the change in the resistance of pure metals at very low temperatures III. The resistance of platinum at helium temperatures", *CPL*, No. 119, 1—26.

——, 1911b, "On the change in the resistance of platinum at helium temperatures", *CPL*, 119b, 1—29.

——, 1911c, "Further experiments with liquid helium C. On the change of electric resistance of pure metals at very low temperature IV. The resistance of pure mercury at helium temperatures", *CPL*, 120a, 3—15.

——, 1911d, "Further experiments with liquid helium, C. On the change of electric resistance of pure metals at very low temperatures, etc. IV. The resistance of pure mercury at helium temperatures", *CPL*, 120b, 17—19.

——, 1911e, "Further experiments with liquid helium, D. On the change of the electrical resistance of pure metals at very low temperatures, etc., V. The disappearance of the resistance of mercury", *CPL*, 122b, 13—15.

——, 1911f, "Further experiments with liquid helium, G. On the electrical resistance of pure metals, etc. VI. On the sudden change in the rate at which the resistance of mercury disappears", *CPL*, 124c, 21—27.

——, 1912, "Sur les résistances éléctriques", *Le théorie du rayonnement et les quanta; Rapports et discussions de la réunion tenue à Bruxelles*, du 30 Octobre au 3 Novembre 1911 sous les auspices de M. E. Solvay, P. Langevin and M. de Broglie, eds., (Paris, 1912), 304—312, reprinted in *CPL*, (*Suppl.*), No 29, 3—11.

——, 1913a, "Further experiments with liquid Helium, H. On the electrical resistance of pure metals etc., VII. The potential difference necessary for the electric current through mercury below 4.19 K", *CPL*, 133a, 3—26.

——, 1913b, "Further experiments with liquid helium H. The potential difference necessary for the electric current through mercury below 4.19 K (Continued)", *CPL*, 133b, and 133c, and 35—48.

——, 1913c, "Further experiments with liquid Helium, H. On the electrical resistance etc. (Continued). VIII. The sudden disappearance of the ordinary resistance of tin, and the superconductive state of lead", *CPL*, 133d, 51—68.

——, 1913d, "Report on the researches made in the Leiden cryogenics laboratory between the second and third international congress of refrigeration", *CPL*. (*Suppl.*) 34b, 37—70.

——, 1913e, "Nobel lecture", reprinted in *Nobel Lectures . . . ; Physics, 1901—1921*, New York: Elsevier, 1967, 306—336.

——, 1914a, "Further experiments with liquid helium, I. The Hall effect, and the magnetic change in resistance at low temperatures, IX. The appearance of galvanic resistance in supra-conductors which are brought into a magnetic field, at the threshold value of the field", *CPL*, 139f, 65—71.

——, 1914b, "Further experiments with liquid helium, J. The imitation of an Ampere molecular current of a permanent magnet by means of a supra-conductor", *CPL*, 140b, 9—18.

——, 1914c, "Further experiments with liquid helium. K. Appearance of beginning paramagnetic saturation", *CPL*, 140d, 29—32.

——, 1914d, "Further experiments with liquid helium. L. The persistence of currents without electromotive force in supra-conducting circuits", *CPL*, 141b, 15—21.

——, 1914e, "Further experiments with liquid helium. M. Preliminary determination of the specific heat and of the thermal conductivity of mercury at temperatures, obtainable with liquid helium, besides some measurements of thermoelectric forces and resistances for the purpose of these investigations", *CPL*, 142c, 25—33.

——, 1917, "Methods and apparatus used in the cryogenic laboratory. XVII. Cryostat for temperatures between 27 °K and 55 °K", *CPL*, 151a.

——, 1919, "Demonstration of liquid helium", *CPL*, (*Suppl.*) 43c, 13—19.

——, 1921, "Les superconducteurs et le modèle de l'atome Rutherford—Bohr", (Rapport presente au troisième Conseil de Physique Solvay, tenu à Bruxelles, Avril 1921), reprinted in *CPL*. (*Suppl.*) No 44, 30—50.

——, 1922, "On the lowest temperature yet obtained", *Faraday Society, Transactions*, **18**, 145—174.

——, 1924, "Rapport sur de nouvelles expériences avec les supraconducteurs". (Fourth International Congress of Refrigeration, London, June 1924, and 4eme Conseil de Physique Solvay, Bruxelles, Avril 1924), *CPL* (*Suppl.*) No 50, 3—34.

——, Beckman, B., 1912a, "On the Hall effect and the change in the resistance in a magnetic field at low temperatures, I. Measurements on the Hall effect and the change in the resistance of metals and alloys in a magnetic field at the boiling point of hydrogen and at lower temperatures", *CPL*, No 129a, 3—16.

——, 1912b, "On the Hall effect, and on the change in resistance in a magnetic field at low temperatures. II. The Hall effect and the resistance increase for bismuth in a magnetic field at, and below, the boiling point of hydrogen", *CPL*, No 129c, 31—34.

——, Boks J. D. A., 1924, "Further experiments with liquid helium, V. The variation of density of liquid helium below the boiling point". (Fourth International Congress of Refrigeration, London, June *1924*), *CPL*, 170b, 18—23.

——, Clay, J., 1906a, "On the measurement of very low temperatures XI. A comparison of the platinum resistance thermometer", *CPL*, 95c, 39—45.

——, 1906b, "On the measurement of very low temperatures XII. A comparison of the platinum resistance thermometer with the gold resistance thermometer", *CPL*, 95d, 49—52.

——, 1907a, "On the change of the resistance of the metals at very low temperatures and the influence exerted on it by small amounts of admixtures", *CPL*, 99c, 17—26.

——, 1907b, "On the measurement of very low temperatures XVI. Calibration of some platinum-resistance thermometers", *CPL*, 99d, 9—14.

——, 1908, "On the change of the resistance of pure metals at very low temperatures and the influence exerted on it by small amounts of admixtures II", *CPL*, 107c, 19—27.

——, Crommelin, C. A., 1910, "Isotherms of monoatomic gases and their binary mixtures, VI. Coexisting liquid and vapour densities of argon. Calculation of the critical density of argon", *CPL*, 118a, 1—10.

——, 1911, "Isotherms of monoatomic substances and of their binary mixtures. IX. The behaviour of argon with regard to the law of corresponding states", *CPL*, 120a, 3—13.

——, 1913, "Isothermals of di-atomic substances and their binary mixtures, XIII. Liquid densities of hydrogen between the boiling point and the triple point; contraction of hydrogen on freezing", *CPL*, 137a, 1—5.

——, 1915, "Isothermals of monoatomic substances and of their binary mixtures, XVII. Isothermals of neon and preliminary determinations concerning the liquid condition of neon", *CPL*, 147d, 47—54.

——, Cath, P. G., 1917a, "Isothermals of diatomic substances and their binary mixtures XVIII. A preliminary determination of the critical point of neon", *CPL*, 151b.
——, 1917b, "Isothermals of diatomic substances and their binary mixtures, XIX. A preliminary determination of the critical point of hydrogen", *CPL*, 151c.
——, De Haas, W. J., 1912, "Isotherms of diatomic substances and of their binary mixtures, XII. The compressibility of hydrogen vapour at and below the boiling point", *CPL*, 127c, 19—33.
——, Dorsman, C., Holst, G., 1914, "Isothermals of diatomic substances and their binary mixtures, XV. Vapour-pressures of oxygen and critical point of oxygen and nitrogen", *CPL*, 145b, 9—15.
——, Francis Hyndman, H. H., 1901, "Isothermals of diatomic gases and their binary mixtures, I. Piezometers of variable volume for low temperatures", *CPL*, 69, 1—10.
——, 1902, "Isothermals of diatomic gases and their binary mixtures, V. An accurate volumenometer and mixing apparatus", *CPL*, 84, 1—9.
——, Keesom, W. H., 1908, "On the equation of state of a substance in the neighbourhood of the critical point liquid-gas I. The disturbance function in the neighbourhood of the critical state", *CPL*, 104a.
——, 1912, "Die Zustandsgleichung" Art. V. 10, *Encyclopädie der mathematischen Wissenschaften*, 615—945.
——, Kuypers, H. A., 1924, "Isotherms of diatomic substances and their binary mixtures, XXV. On the isotherms of oxygen at low temperatures", *CPL*, 169a, 1—9.
——, Oosterhuis, E., 1912, "Magnetic Researches VI. On paramagnetism at low temperatures", *CPL*, 129b, 19—27.
——, 1913a, "Magnetic Researches VI. On paramagnetism at low temperatures (continued)", *CPL*, 132e, 45—52.
——, 1913b, "Magnetic researches. VIII. On the succeptibility of gaseous oxygen at low temperatures", *CPL*, 134d, 31—33.
——, Perrier, A., 1910, "Researches on the magnetization of liquid and solid oxygen", *CPL*, 116, 1—43.
——, 1911a, "Researches on magnetism III. On the para- and dia-magnetism at very low temperatures", *CPL*, 122a, 1—10.
——, 1911b, "Researches on magnetism IV. On paramagnetism at very low temperatures", *CPL*, 124a, 1—8.
——, Tuyn, W., 1922, "Further experiments with liquid helium R. On the electrical resistance of pure metals etc. XI. Measurements concerning the electrical resistance of ordinary lead and uranium lead below 14 °K", *CPL*, 160b.
——, Van Urk, A. Th., 1924, "Isotherms of diatomic substances and their binary mixtures XXVIII. On the isotherms of nitrogen at low temperatures", *CPL*, 169e, 45—60.
——, and Weber, S., 1913a, "Investigation of the viscosity of gases at low temperatures III. Comparison of the results obtained with the law of corresponding states", *CPL*, 134c, 24—27.
——, 1913b, "Vapour pressures of substances of low critical temperature at low reduced temperatures I. Vapour pressures of carbon dioxide between −160°C and −183°C", *CPL*, 137b, 7—23.
——, Zakrzewski, C., 1904, "Contribution to the knowledge of Van der Waals' ψ-surface IX. The conditions of coexistence of binary mixtures of normal substances according to the law of corresponding states", *CPL* (*suppl.*), 8.
Kapitza, P., 1938, "Viscosity of liquid helium below the λ-point", *Nature*, **141**, 74.
——, 1941a, "The study of heat transfer in helium II", *Journal of Physics* (USSR), **4**, 181. Reprinted in D. Ter Haar, 1965, Vol. II, 581—624.
——, 1941b, "Heat transfer and superfluidity of helium II", *Journal of Physics* (USSR), **5**, 59. Reprinted in D. Ter Haar 1965, Vol. II, 625—639.
——, 1941c, "Problems of liquid helium", A report at the General Assembly of the USSR Academy of Sciences, 28 Dec. 1940, translated from *Sovetskaya Nauka*, **1**, 33, in P. L. Kapitza, 1980, 12—34.
——, 1980, *Experiment, Theory, Practice. Articles and Addresses*, Boston: Reidel.
Keesom, W. H., 1901, "Contributions to the knowledge of van der Waals ψ-surface V. The dependence of the plait point constants on the composition in binary mixtures with small

proportions of one of the components", *CPL*, 75.
——, 1912a, "On the deduction of the equation of state from Boltzmann's entropy principle", *CPL* (*suppl.*), 24a, 1—20.
——, 1912b, "On the deduction from Boltzmann's entropy principle of the second virial coefficient for material particles (in the limit rigid spheres of central symmetry) which exert central forces upon each other and for rigid spheres of central symmetry containing an electrical doublet at their centre", *CPL* (*suppl.*), 24b, 21—41.
——, 1912c, "On the second virial coefficient for diatomic gases", *CPL* (*suppl.*), 25.
——, 1912d, "On the second virial coefficient for monoatomic gases, and for hydrogen below the Boyle-point", *CPL* (*suppl.*), 26.
——, 1915a, "The second virial coefficient for rigid spherical molecules, whose mutual attraction is equivalent to that of a quadruplet placed at their center", *CPL* (*suppl.*), 39a.
——, 1915b, "Two theorems concerning the second virial coefficient for rigid spherical molecules which besides collisional forces only exert Coulomb forces and for which the total change of the active agent is zero", *CPL* (*suppl.*), 39b.
——, 1926a, "Prof. Dr. H. Kamerlingh Onnes. His life-work, the founding of the Cryogenic Laboratory", *CPL* (*suppl.*), 57, 3—21.
——, 1926b, "Solid helium", *CPL*, 184b, 9—20.
——, 1926c, "Solidification of helium", *Nature*, **118**, 81.
——, 1927, "The melting-curve of Helium and the heat-theorem of Nernst", *CPL* (*suppl.*), 61b, 9—15.
——, 1928, "The states of aggregation of condensed helium", *Nature*, **122**, 847—849.
——, 1932, "Quelques remarques en rapport avec l'anomalie de la chaleur spécifique de l'hélium liquide au point lambda", [6e Congres International du Froid, Buenos Aires, Août-Septembre 1932], *CPL* (*suppl.*), 71e, 47—52.
——, 1933, "On the jump in the expansion coefficient of liquid helium in passing the lambda-point", *CPL* (*suppl.*), 75a, 1—7.
——, 1935, "Lambda-phenomena in liquid helium", *Proceedings of the Royal Society, London*, **A152**, 11—13.
——, 1942, "*Helium*", Amsterdam: Elsevier.
——, Clusius, K., 1932, "Ueber die spezifische Wärme des flussigen Heliums", *CPL*, 219e, 42—58.
——, Kamerlingh Onnes, H., 1924, "On the question of the possibility of a polymorphic change at the point of transition into the supraconductive state", *CPL*, 174b, 43—45.
——, Keesom, A. P., 1932, "On the anomaly in the specific heat of liquid helium", *CPL*, 221d, 19—26.
——, 1935, "New measurements on the specific heat of liquid helium", *Physica*, **2**, 557—569.
——, 1936, "On the heat conductivity of liquid helium", *Physica*, **3**, 359—360.
——, Saris, B. F., 1938, "A few measurements on the heat conductivity of liquid helium II", *Physica*, **5**, 281—285.
——, Kok, J. A., 1932, "On the change of the specific heat of tin when becoming supraconductive", *CPL*, 221e, 27—32.
——, MacWood, G. E., 1938, "The viscosity of liquid helium", *Physica*, **5**, 737—744.
——, Taconis, K. W., 1938a, "On the structure of solid helium", *Physica*, **5**, 161—169.
——, 1938b, "Debye—Scherrer exposures of liquid helium", *Physica*, **5**, 270—280.
——, van der Ende J. N., 1930, "The specific heat of solid substances at the temperatures obtainable with the aid of liquid helium II. Measurements of the atomic heats of lead and of bismuth", *Koninklijke Academie van wetenschappen te Amsterdam, Proceeding of the section of Sciences*, **33**, 243—254.
——, 1932, "The specific heat of solids at temperatures obtainable with liquid helium IV. Measurements of the atomic heats of tin and zinc", *CPL*, 219b, 10—25.
——, Van Leeuwen, C., 1915, "On the second virial coefficient for rigid spherical molecules carrying gudruplets", *CPL* (*suppl.*), No 39c.
——, Wolfke, M., 1927, "Two different liquid states of helium", *CPL*, 190b, 17—22.

[Kelvin, Lord] Thomson, W., 1902, "Aepinus atomized", *Philosophical Magazine*, **3**, 257—283.

Kikoin, A. K., Lasarew, B. G., 1938, "Experiments with liquid helium II", *Nature,* **141**, 912—913.

Klein, M. J., 1970, *Paul Ehrenfest*, Vol. 1, "The making of a theoretical physicist", Amsterdam-London: North-Holland (Esp. Chapter 1: The new professor at Leyden, pp. 1—16, and Chapter 9: Early years in Leyden, pp. 193—216).

——, 1974, "The historical origins of the van der Waals equation", *Physica*, **73**, 28—47.

Koppe, H., 1950, "Theorie der Supraleitung", *Ergebnisse der exakten Naturwissenschaften*, **23**, 283—358.

Koopman, J. F. H., 1927, "The late professor dr. H. Kamerlingh Onnes", *Cold Storage*, **XXIX**, No 336, 129—130.

Kronig, R., 1932, "Zur theorie der Superleitfähigheit", *Zeitschrift für Physik*, **78**, 744—750.

——, 1933, "Zur theorie der Superleitfähigheit II", *Zeitschrift für Physik*, **80**, 203—216.

——, 1935, "The propagation of electromagnetic waves in metallic conductors and its bearing upon the problem of supraconductivity", *Proceedings of the Royal Society of London*, **A152**, 16—19.

Kuhn, T. S., 1962a, "The historical structure of scientific discovery", *Science*, **136**, 760—764; Reprinted in T. S. Kuhn, 1977, 165—177.

——, 1962b, *The Structure of Scientific Revolutions*, Chicago: Chicago University Press. (Second edition, enlarged, 1970).

——, 1977, *The Essential Tension*, Chicago: Chicago University Press.

Kurti, N., "The temperature range below 1° absolute", in F. Simon *et al.*, 1952, 30—65.

——, 1958, "Franz Eugen Simon", *Biographical Memoirs of Fellows of the Royal Society*, **4**, 225—256.

Labadie, J., 1932, "La physique au voisinage du zéro absolu; une visite au laboratoire de l'université de Leyde (Pays-Bas)", *La Science et le Vie*, **XLI**, No 180, 439—450.

Lakatos, I., 1978, *Philosophical Papers*, Vol. I and II, edited by J. Worral and G. Currie, Cambridge: Cambridge University Press.

——, Musgrave, A., 1970, eds., *Criticism and the Growth of Knowledge*, Cambridge: Cambridge University Press.

Landau, L. D., 1937a, "K teorii fazovnykh perekhodov", *Zhurnal eksperimentalnoi i teoreticheskoi fiziki*, **7**, 19—32. "On the theory of phase transitions", in D. Ter Haar, 1965.

——, 1937b, "K teorii sverkhpovodnimosti", *Zhurnal eksperimentalnoi i teoreticheskoi fiziki*, **7**, 371. "On the theory of superconductivity", in D. Ter Haar, 1965, 217—225.

——, 1938, "The intermediate state of supraconductors", *Nature*, **141**, 688. Reprinted in D. Ter Haar, 1965, 266—267.

——, 1941, "The theory of superfluidity of helium II", *Journal of Physics* (USSR), **5**, 71—90. Reprinted in Galasiewicz, 1971,

——, 1943, "On the theory of the intermediate state of superconductors", *Journal of Physics* (USSR), **7**, 99—107. Reprinted in D. Ter Haar, 1965, 365—373.

——, 1944, "On the hydrodynamics of helium II", *Journal of Physics* (USSR), **VIII**, No 1, 1—3.

——, 1947, "On the theory of superfluidity of helium II", *Journal of Physics* (USSR), **11**, 9—92. Reprinted in Galasiewicz, 1971, 243—246.

——, 1949, "On the theory of superfluidity", *Physical Review*, ser. 2, **75**, 884—885. Reprinted in D. Ter Haar, 1965, 474—477.

——, 1956, "The theory of a Fermi liquid", *Zhurnal eksperimentalnoi i teoreticheskoi fiziki*, **30**, 1058—1064.

——, 1957, "Oscillation in a Fermi liquid", *Zhurnal eksperimentalnoi i teoreticheskoi fiziki*, **32**, 59—66.

——, Ginzburg, V. L., 1950, "K teorii sverkhprovodnimosti", *Zhurnal eksperimentalnoi i teoreticheskoi fiziki*, **20**, 1064. "On the theory of superconductivity", in D. Ter Haar, 1965, 546—568.

——, Khalatnikov, I. M., 1949, "The theory of the viscosity of helium II. 1. Collisions of elementary excitations in helium II", *Journal of Experimental and Theoretical Physics* (USSR), **19**, 637. Reprinted in D. Ter Haar, 1965, 511—531.

Lane, C. T., 1962, *Superfluid Physics*, New York: McGraw-Hill.

——, 1967, "Resource letter LH-1 on liquid He", *American Journal of Physics*, **235**, 367—375.

Langevin, P., de Broglie, L., eds., 1912, *La théorie du rayonnement et les quanta; Rapports et discussions de la réunion tenue à Bruxelles du 30 Octobre au 3 Novembre 1911 sous les auspices de M. E. Solvay*, Paris: Gautier-Villars.

Laudan, L., 1977, *Progress and its Problems*, Berkeley: University of California Press.

——, 1980, "Why was the logic of discovery abandoned?", in T. Nickles, 1980a, 173—184.

——, Donovan, A., Laudan, R., Barker, P., Brown, H., Leplin, J., Thagard, P., Wykstra, S., 1986, "Scientific change: philosophical models and historical research", *Synthese*, **69**, 141—223.

Lazarides, G., Shafi, Q. I., (eds.), 1986, *Particles and the Universe*, Amsterdam: North Holland Physics Publishing.

Lemaire, E., 1911, "L'organization du laboratoire cryogène de Leyde (Pays-Bas). Dernières recherches du professeur Kamerlingh Onnes sur la resistivité élétrique des metaux purs au voisinage du zéro absolu, *La Révue generale du Froid*, **III**, No 10, 475—483.

Leprince, F., 1974, "1913—1973": Soixante ans de solenoïdes supraconducteurs", *Révue des questions scientifiques*, **145**, 233—254 and 537—564.

Levelt Sengers, J. M. H., 1974, "From van der Waals equation to the scaling laws", *Physica*, **73**, 73—106.

Lifshitz, E., 1944, "Radiation of sound in helium II", *Journal of Physics* (USSR), **8**, 110—114. Reprinted in Z. M. Galasiewicz 1971, 234—242.

——, 1958, "Superfluidity", *Scientific American*, **198**(6), 30—35.

——, Andronikashvili, E. L., 1959, *A supplement to* (*W. H. Keesom's*) '*Helium*', translated from Russian, New York: Consultants Bureau Inc.

Lindemann, F. A., 1911, "Untersuchungen über die spezifische Wärme bei tiefen temperaturen, IV" *Sitzungsberichte, Akademie der Wissenschaften*, 316—321.

——, 1915, "Note on the theory of the metallic state", *Philosophical Magazine*, **29**, 127—140.

Lippmann, G., 1919, "Sur les propriétés des circuits electriques denues de résistance", *Académie des Sciences, Comptes Rendus*, **168**, 73—78.

London, E., 1961a, "Fritz London: A brief biography", in F. London 1961, Vol. I, X—XIV.

——, 1961b, "A list of publications of Fritz London", in F. London, 1961, Vol. I, XV—XVIII.

London, F., 1935, "Macroscopical interpretation of superconductivity", *Proceedings of the Royal Society of London*, **A152**, 24—34.

——, 1936, "On condensed helium at absolute zero", *Proceedings of the Royal Society of London*, **A153**, 576—586.

——, 1937, Une conception nouvelle de la supraconductibilite, *Actualités Scientifiques et Intustrielles*, **458**, 5—80.

——, 1938a, "The λ-point phenomenon of liquid helium and the Bose-Einstein degeneracy", *Nature*, **141**, 643—644.

——, 1938b, "On the Bose—Einstein condensation", *Physical Review*, ser. 2, **54**, 947—954.

——, 1939, "The state of liquid He near absolute zero", *Journal of Physical Chemistry*, **43**, 49—69.

——, 1945, "Planck's constant and low temperature transfer", *Reviews of Modern Physics*, **17**, 310—320.

——, 1947, "The present state of the theory of liquid helium", International Conference on Fundamental Particles and Low Temperatures, Cavendish Laboratory, Cambridge 22—27 July 1946. Published by the Physical Society, Vol. II, 1—18.

——, 1948, "On the problem of the molecular theory of superconductivity", *Physical Review*, **74**, 562—573.

——, 1949a, "Program for the molecular theory of superconductivity", *Proceedings of the International Conference on the Physics of very Low Temperatures*, Cambridge Massachussetts, (September 6—10), 76—83.

——, 1949b, "The rare isotope of helium, He^3; A key to the strange properties of ordinary liquid helium, He^4", *Nature*, **163**, 694—696.

——, 1950, *Superfluids*, New York: Willey.

——, 1951, "Limitations of the two-fluid theory", *Proceedings of the Royal Society of London, Low Temperature Physics, August 22—28*, (R. Bowers ed.), 2—6.

——, 1961, *Superfluids*, 2 Vols, 2nd edition, New York: Dover.
——, London, H., 1935, "The electromagnetic equations of the supraconductor", *Proceedings of the Royal Society of London*, **A149**, 71—88.
——, Rice, O., 1948, "On solutions of He^3 in He^4", *Physical Review*, ser. 2, **73**, 1188—1193.
London, H., 1939, "Thermodynamics of the thermomechanical effect of liquid He II", *Proceedings of the Royal Society* (London), **A171**, 484—496.
——, 1960, "Superfluid He", (First Simon memorial lecture), *Yearbook of the Physical Society* (London), 34—38.
Lorentz, H. A., 1905, "The motion of electrons in metallic bodies", *Proceedings, Akademie der Wetenschappen, Amsterdam*, **7**, 438—453, 585—593, 684—691.
——, 1908, "La liquefaction de l'hélium", *Archives Neerlandaises des Sciences exactes et naturelles*, serie II, Tome XIII, 492—502.
——, 1924, "Application de la théorie des electrons aux proprietés des metaux", *Rapports et Discussions du Quatrième Conseil de Physique Solvay*, Preprinted in *CPL* (*suppl.*), 50b.
Luttinger, J. M., 1950, "A note on Tisza's theory of superconductivity", *Physical Review*, **80**, No 4, 727—729.
MacDonald, D. K. C., "Electrical conductivity in metals and alloys at low temperatures", in *Handbuch der Physik*, S. Flugge, ed., Vol. 14, 137—197.
MacWood, G. E., 1938a, "The theory of the measurements of viscosity and slip of fluids by the oscillating disk method", *Physica*, **5**, 374—384.
——, 1938b, "The theory of the measurements of viscosity and slip of fluids by the oscillating disc method, part II", *Physica*, **5**, 763—768.
Marshak, R. E., ed., 1966, *Perspectives in modern physics* (Essays in honour of Hans Bethe), New York.
Martinez, J. P., Kamerlingh Onnes, H., 1923, "Isothermes des substances monoatomiques et de leurs mélanges binaires. XXI. Idem substances diatomiques XXI. Determinations d'isothermes de l'hydrogène et de l'hélium a basse temperature, faites en vue d'examiner si la compressibilité de ces gaz est influencé par les quanta", *CPL*, No 164, 1—26.
Mathias, M. E., 1894, "Travaux Recents sur la Continuite des Etas Gageaus et Liquide et sur la Notion Generalisée d'Etas Correspondents" in the French translation of J. D. van der Waals' Thesis *La Continuite des Etats Gageux et Liquide*, Paris, 270—279.
——, 1926, "H. Kamerlingh Onnes (1853—1926). L'oeuvre et l'homme", *Revue generale des sciences*, **37**, 294—298.
——, 1928, "Le diamètre rectiligne de la courbe des densités comme point de départ d'une classification naturelle des corps purs, simples, ou composés", *CPL* (*suppl.*), No 64b.
Maxwell, E., 1950, "Isotope effect in the superconductivity of mercury", *Physical Review*, **78**, 477.
McLennan, J. C., 1934, "Electrical phenomena at extremely low temperatures", *Reports on Progress in Physics*, (Cambridge, 1934), **1**, 198—227.
——, 1935, "Opening address in a discussion on supraconductivity and other low temperature phenomena", *Proceedings of the Royal Society of London*, **A152**, 1—8.
——, Smith, H. D., Wilhelm, J. O., 1932, "The scattering of light by liquid helium", *Philosophical Magazine*, ser. 7, **14**, 161—167.
Mehra, J., ed., 1973, *The Physicist's Conception of Nature*, Dordrect: Reidel.
Meissner, W., 1935, "The magnetic effects occuring on transition to the supraconductive state", *Proceedings of the Royal Society* (London), **A152**, 13—15.
——, Ochsenfeld, R., 1933, "Ein neuer Effekt bei Eintritt der Supraleitfähigkeit", *Die Naturwissenschaften*, **21**, 787—788.
Mendelssohn, K., 1945, "The frictionless state or aggregation", *Proceedings of the Physical Society* (London), **57**, Part 5, No 323, 371—389.
——, 1946, "Superconductivity", *Reports of Progress in Physics*, **X**, 358—377.
——, 1949, "Low temperature physics", *Reports of Progress in Physics*, **XII**, 270—290.
——, 1952, "Superconductivity", in F. Simon *et al.*, 1952, 95—132.
——, 1956, "Liquid Helium" in S. Flugge (ed.), *Handbuch der Physic*, **XV**, 371—461.
——, 1963, "Physical Sciences in the twentieth century", *Contemporary Physics*, **4**, 422—434.
——, 1964, "Prewar work on superconductivity as seen from Oxford", *Reviews of Modern Physics*,

36, 7—12.
——, 1973, "Superconductivity and Superfluidity", in H. Hagen, M. Wagner, eds., *Cooperative phenomena*, Berlin: Springer 426—435.
——, 1977, *The Quest for Absolute Zero*, Second Edition with S. I. Units, London: Taylor and Francis.
——, Babbitt, J. D., 1934, "Persistent currents in Supraconductors", *Nature*, **133**, 459—460.
Mermin, N. D., Lee, D. M., 1976, "Superfluid helium 3", *Scientific American*, **235**(6), 56—71.
Moulines, C. U., 1976, "Approximate application of empirical theories: A general explication", *Erkentnis*, **10**, 201—223.
Musgrave, A., 1976, "Method or Madness" in R. S. Cohen *et al.*, 1976, 457—492.
Nernst, W., 1906, "Ueber die Berechnung chemischer Gleinchgewichte aus thermischen Messungen", *Nachrichten von der Königlichen Gesellschaft der Wissenschaften zu Göttingen*, **1**, 1—40.
——, 1911, "Untersuchungen über die spezifische Wärme bei tiefen Temperaturen, III", *Sitzungsberichte, Akademie der Wissenschaften*, 306—315.
——, 1912, "Application de la théorie des quanta à divers problèmes physicochimiques", in P. Langevin, M. de Broglie, 1912, 254—290.
——, 1918, *Die Thoretische und Experimentellen Grundlagen des neuen Wärmesatzes*, Halle: Knapp.
——, Lindemann, F. A., 1911, "Spezifische Wärme und Quanten-theorie", *Zeitschrift fur Elektrochemie*, **17**, 817—827.
Nersessian, N. J., 1984, *Faraday to Einstein: Constructing Meaning in Scientific Theories*, Dordrecht: Martinus Nijhoff.
Nickles, T., 1978, "Scientific problems and constraints", in P. Asquith and I. Hacking, (eds.), *PSA 1978*, 134—148, East Lansing, Mich.: Philosophy of Science Association.
——, ed., 1980a, *Scientific Discovery, Logic and Rationality*, Dordrecht: Reidel.
——, 1980b, *Scientific Discovery: Case studies*, Dordrecht: Reidel.
——, 1980c, "Scientific discovery and the future of philosophy of science" in T. Nickles, 1980a, 1—60.
——, 1980d, "Can scientific constraints be violated rationally?", in T. Nickles, 1980a, 285—316.
——, 1981, "What is a problem that we may solve it?", *Synthese*, **49**, 85—118.
——, forthcoming, "Questioning and problems in philosophy of science", in M. Meyer, W. de Gruyter, eds., *Questions and Questioning.*
Nordheim, L. W., 1961, "Fritz London: 1900—1954, an appreciation of his work", in F. London, 1961, Vol. I, V—VIII.
Ogg, R., 1946, "Bose—Einstein condensation of trapped electron pairs. Phase separation and superconductivity of metalammonia solutions", *Physical Review*, ser. 2, 243—244.
Olszewski, C., 1895, "On the liquefaction of gases", *Philosophical Magazine*, **39**, 188—212.
Osheroff, D., Richardson, R., Lee, D., 1972, "Evidence for a new phase of solid ^{3}He", *Physical Review Letters*, **28**, 885.
Parks, R. D., (ed.), 1969, *Superconductivity*, 2 Vols., New York: Marcel Dekker.
Pellam, J. P., Scott, R. B., 1949, "Second sound velocity in paramagnetically cooled liquid helium II", *Physical Review*, ser. 2, 76, 869—870.
Perrier, A., Kamerlingh Onnes, H., 1914a, "Magnetic researches. XII. The susceptibility of solid oxygen in two forms", *CPL*, No 139c, 28—34.
——, 1914b, "Magnetic investigations XIII. The susceptibility of liquid mixtures of oxygen and nitrogen and the influence of the mutual distance of the molecules upon paramagnetism", *CPL*, 139d, 37—54.
Peshkov, V., 1944, "'Second sound', in helium II", *Journal of Physics* (USSR), **8**, 381—389.
——, 1946, "Determination of the velocity of propagation of the second sound in helium II", *Journal of Physics* (USSR), **10**, 389—398. Reprinted in Z. Galasiewitz, 1971, 191—233.
——, 1948, "Skorost' vtorogo zvuka ot 1.3 do 1. pe°K", *Zhurnal eksperimentalnoi i teoreticheskoi fiziki*, **18**, 951—952.
Picard, E., 1939, "Les basses températures et l'oeuvre de M. Kamerlingh Onnes", *Academie des Sciences, Paris, Mémoires*, **63**, 1—24.

Pines, D., 1958, "Superconductivity in the Periodic system", *Physical Review*, **109**, 280—287.

Pippard, A. B., 1953, "An experimental and theoretical study of the relation between magnetic field and current in superconductor", *Proceedings of the Royal Society* (London), **A216**, 547—568.

——, 1964, "Unsolved problems of superconductivity", *Reviews of Modern Physics*, **36**, 328—331.

——, 1986, "Early superconductivity research (except Leiden)", paper presented at the *H. Kamerlingh Onnes Symposium on the Origins of Applied Superconductivity*, October 1, Maryland, U.S.A.

Planck, M., 1911a, "Eine neue strahlungshypothese", *Verhandlungen der Deutsche Physikalische Gesellschaft*, **13**, 138—148.

——, 1911b, *Thermodynamik*, Leipzig: Verlag von Veit.

——, 1912, "La loi du rayonnement noir", in P. Langevin, M. de Broglie, 1912, 93—114.

Popper, K. R., 1959, *Logic of Scientific Discovery*, London: Hutchinson.

——, 1962, *Conjectures and Refutations*, New York: Basic Books.

——, 1972, *Objective Knowledge*, Oxford: Oxford University Press.

Prigogine, I., Mazur, P., 1951, "Sur deux Formulations de l'Hydrodynamique et le Probleme de l'Helium Liquide II" *Physica*, **17**, 661—679.

Rayfield G., Reif, F., 1963, "Evidence for the creation and motion of quantized vortex rings in superfluid helium", *Physical Review Letters*, **11**, 305—308.

Reif, F., 1960, "Superfluidity and 'Quasi-Particles'", *Scientific American*, **230**(5), 138—150.

——, 1968, "Superfluidity: the paradox of atomic simplicity and remarkable behaviour", in N. Wiser, D. J. Amit, 1970, 1—14.

Reynolds, C. A., Serin, B., Wright, W. H., Nesbitt, L. B., 1950, "Superconductivity of isotopes of mercury", *Physical Review*, **78**, 487.

Richardson, M. O. W., 1924, "Encore une théorie de la conductibilité métallique", in *Conductibilite électrique des métaux et problèmes connexés, rapports et discussion du 4ème conseil Solvay*, (Avril 1924), (Paris 1927), 135—146.

Rickayzen, C., 1965, *Theory of Superconductivity*, New York: Interscience.

Riecke, E., 1898, "Zur Theorie des Galvanismus und der Wärme" *Annalen der Physik*, **66**, 353—389, and 541—581.

——, 1900, "Über das Verhaltnis der Leitfähigkeiten der Metalle für Wärme und fur Elektricität", *Annalen der Physik*, **2**, 835—842.

Rollin, B. V., 1936, "Investigations relating to the thermal insulation of vessels containing liquid helium at temperatures below the λ point", *Actes du VIIe Congrès International du Froid, La Haye: 16—19 Juin 1936*, and *Amsterdam: 20 Juin 1936*, Utrecht, 1937, Vol. I, 187—189.

Ruhemann, M., Ruhemann, B., 1937, *Low Temperature Physics*, Cambridge: Cambridge University Press.

Rutgers, A. J., 1934, "Note on supraconductivity", *Physica*, **1**, 306—320.

Satterly, J., 1936, "The physical properties of solid and liquid helium", *Reviews of Modern Physics*, **8**, 347—357.

Schilpp, P. A., ed., 1974, *The Philosophy of Karl Popper*, La Salle, Illinois: Open Court.

Schrieffer, J. R., 1973, "Macroscopic quantum phenomena from pairing in superconductors", *Physics Today*, **26**(7), 23—28.

Serin, B., 1956, "Superconductivity. Experimental part", in S. Fluge, ed., *Encyclopedia of Physics*, Berlin: Springer, Vol. XV, 210—273.

——, Reynolds, C. A., Nesbitt, L. B., 1950, "Mass dependence of the superconducting transition temperature of mercury", *Physical Review*, **80**, 761—762.

Schafroth, M. R., 1951, "A note on Fröhlich's theory of superconductivity", in R. Powers, ed., *Proceedings of the International Conference on Low Temperature Physics*, held in Oxford, August 22—28, 112—113.

——, 1960, "Theoretical aspects of superconductivity", *Solid State Physics*, **10**, 293—498.

Shapere, D., 1980, "The character of scientific change", in T. Nickles, 19801, 61—101.

Shoenberg, D., 1952a, *Superconductivity*, Cambridge: Cambridge University Press.

——, 1952b, "Application of second quantization methods to the classical statistical mechanics (I)", *Nuovo Cimento*, **9**, 1139.

——, 1953, "Application of second quantization methods to the classical statistical mechanics (II)", *Nuovo Cimento*, **10**, 419—472.

——, 1971, "Heinz London", *Royal Society of London, Biographical Memoirs*, **17**, 441—461.

——, 1978, "Forty odd years in the cold. Reminiscences of work in low temperature physics", *Physics Bulletin*, **29**(1), 16—19.

Silsbee, F. B., 1916, "Electrical conduction in metals at low temperatures", *Journal of the Washington Academy of Sciences* **VI**, 597—602.

——, 1917, *Scientific papers of the Bureau of Standards*, **14**, No 307.

Simon, F., 1927, "Zum Prinzip von der Unerreichbarkeit des absoluten Nullpunktes", *Zeitschrift für Physik*, **41**, 806—809.

——, 1930, "Funfundzwanzig Jahre Nernstscher Warmesatz", *Erg. ex. Naturw*, **9**, 222—274.

——, 1931, "Uber den Zustand der unterkühlten Flussigkeiten und Gläser", *Zeitschrift für anorganische und allgemeine Chimie*, **203**, 219—227.

——, 1934, "Behaviour of condensed helium near absolute zero", *Nature*, **133**, 529.

——, 1952, "Low Temperature problem, a general survey", in F. Simon *et al.*, 1952, 1—29.

——, and Lange, F., 1926, "Zur Frage der Entropie amorpher Substanzen", *Zeitschrift für Physik*, **38**, 227—236.

——, *et al.*, 1952, *Low Temperature Physics, Four Lectures*, New York: Academic Press.

Simon, H., 1977, *Models of Discovery and Other Topics in the Methods of Science*, Dordrecht: Reidel.

Sizoo, G. J., de Haas, W. J., and Kamerlingh Onnes, H., 1926, "Measurements on the magnetic disturbance of the supraconductivity with tin I. Influence of elastic deformation. II. Hysteresis phenomena", *CPL*, 180c, 29—35.

——, Kamerlingh Onnes, H., 1925, "Further experiments with liquid helium CB. Influence of elastic deformation on the supraconductivity of tin and indium", *CPL*, 180b, 14—26.

Slater, J. C., 1934, "Electronic Structure of Matter", *Reviews of Modern Physics*, **6**, 208—280.

Sneed, J., 1988, "Machine models for the growth of knowledge: Theory nets in PROLOG", in K. Gavroglu, Y. Goudaroulis, P. Nikolacopoulos, 1988.

Somerfeld, A., 1928, "Zur Elekronentheorie der Metalle auf Grund der Fermischen Statistik", *Zeitschrift für Physik*, **47**, 1—32.

Sydoriak, S., Grilly, E., Hammel, E., 1949, "Condensation of pure He^3 and its vapour pressures between 1.2° and its critical point", *Physical Review*, ser. 2, **75**, 303—305.

Ter Haar, ed., 1965a, *Collected Papers of P. L. Kapitza*, London: Pergamon Press.

——, 1965b, *Collected Papers of L. D. Landau*, London: Pergamon Press.

Thomson, Sir G., 1958, "Frederick Alexander Lindemann, Viscount Cherwell", *Biographical Memoirs of Fellows of the Royal Society*, **4**, 45—71.

Thomson, J. J., 1915, "Conduction of electricity through metals", *Philosophical Magazine*, **30**, 192—202.

Tisza, L., 1938a, "Transport Phenomena in helium II", *Nature*, **141**, 913.

——, 1938b, "Sur la supraconductibilité thermique de l'hélium II liquide et le statistique de Bose—Einstein", *Comptes Rendus hebdomadaires des Séances de l'Académie des Sciences*, Paris, **207**, 1035—1037.

——, 1938c, "La viscocité de l'helium liquide et la statistique de Bose-Eintein", *Comples Rendus hebdomadaires des Seances de l'Academie des Sciences*, Paris, **207**, 1186—1189.

——, 1940a, "Sur la theorie des liquides quantiques. Application a l'helium liquide. I", *Journal de Physique et Radium*, **1**, 164—172.

——, 1940b, "Sura la theorie des liquides quantiques. Application a l'helium liquide. II", *Journal de Physique et Radium*, **1**, 350—358.

——, 1947, "The theory of liquid helium", *Physical Review*, ser. 2, **72**, 838—854.

——, 1948, "Helium the unruly liquid", *Physics Today*, **1**(4), 4—8 and 26.

——, 1949a, "On the theory of Superfluidity", *Physical Review*, **75**, 885—886.

——, 1949b, "The present state of the helium problem", *Proceedings of the International conference on the Physics of very low temperatures, Cambridge, Massachusetts, Sept. 6—10*, 1—3.

——, 1950, "Theory of superconductivity", *Physical Review*, No 4, 717—726.

——, 1951, "Theory of superconductivity", R. Powers, ed., Proceedings of the International Conference on Low Temperature Physics, held in Oxford, August 22—28, 113—114.

Toulmin, S., 1974, "Scientific strategies and historical change", in K. Schaffner, R. S. Cohen, (eds.), *PSA 1972*, 401—414, Dordrecht: Reidel.

Trigg, G. L., 1975, *Landmark Experiments in 20th century Physics*, New York: Crane, Russak and Co.

Tuyn, W., 1929, "Quelques essais sur les courants persistants", *CPL*, 198, 3—18.

——, Kamerlingh Onnes, H., 1926, "Further experiments with liquid Helium, AA. The disturbance of supra-conductivity by magnetic fields and currents. The hypothesis of Silsbee", *CPL*, 174a, 3—39.

Uhlenbeck, G. E., 1927, "*Over statistische methoden in de theorie der quanta*",'s Gravenhage.

Van Agt, F.P.G.A.J., 1925, "Isotherms of diatomic substances and their binary mixtures, XXXII. On the behaviour of hydrogen according to the law of corresponding states", *CPL*, No 176c,

Van der Waals, J. D., 1880, "Théorie thermodynamique de la capillarité dans l'hypothèse d'une variation continue de densité", *Archives Neerlandaises*, **XXVIII**.

——, 1888, *The Continuity of the Liquid and Gaseous States*, Physical Memoirs of the Physical Society of London, London (translation from the German of J. D. van der Waals' thesis, published originally in Dutch in 1873). Translated by R. Threlfall, F. Adair.

——, 1890, "Theorie moleculaire d'une substance composée de deux matières", *Archives Neerlandaises*, **24**, 1—56.

——, 1910, *Nobel Lectures in Physics 1901—1921*, Elsevier Publishing Co: New York, 1967.

——, Ph. Kohnstamm, 1912, *Lehrbuch der Thermodynamik*, Leipzig.

Van Laer, P. H., Keesom, W. H., 1938, "On the reversibility of the transition process between the superconductivity and the normal state", *Physica*, **5**, 993—997.

Van Urk, A. Th., Keesom, W. H., Kamerlingh Onnes, H., 1925, "Measurements of the surface tension of liquid helium", *CPL*, 179a, 3—8.

Vinen, W., 1958, "Detection of single quanta of circulation in rotating helium II", *Nature*, **181**, 1524—1525.

——, 1961, "The detection of single quanta of circulation in liquid helium II", *Proceedings of the Royal Society* (London), **A260**, 218—236.

Von Laue, M., 1932, "Zur Deutung einger Versuche über Supraleitung", *Physikalische Zeitschrift*, **33**, 793—975.

——, 1949, *Theorie der Supraleitung*, 2nd edition, Berlin: Springer.

——, 1952, *Theory of Superconductivity*, (translated by Meyer and Band), New York: Academic Press.

Watkins, J. W. N., 1970, "Against normal science", in Lakatos I. and Musgrave, A., 1970, 25—38.

Weinberg, S., 1980, "Conceptual foundations of the unified theory of weak and electromagnetic interactions", *Reviews of Modern Physics*, **52**, 515—523.

Weiss, P., 1907, "L'hypothèse du champ coléculaire et la propriété ferromagnetique", *Journal de Physique*, 4e serie, **VI**, 661—690.

——, 1910, "Measure de l'intensité d'animantation à saturation en valeur absolu", *Journal de Physique*, 4e serie, **IX**, 373—393.

——, Kamerlingh Onnes, H., 1910, "Researches on magnetization at very low temperatures", *CPL*, 114, 1—34.

Welker, H., 1938, "Über ein elektronentheoretisches Modell des Supraleiters", *Physikalische Zeitschrift*, **39**, 920—925.

Wilhelm, J. D., Misener, A. D., Clark, A. R., 1935, "The viscosity of liquid helium", *Proceedings of the Royal Society of London*, **A151**, 342—347.

Wilson, A. H., 1931, "The theory of electronic semi-conductors", *Proceedings of the Royal Society* (London), **A133**, 458—491.

——, 1936a, "Superconductivity and the theory of metals", *Reports on Progress in Physics*, **III**, 262—271.

——, 1936b, (2nd ed. 1953), *Theory of Metals*, Cambridge: Cambridge University Press.

Wiser, N., Amit, D. J., eds., 1970, *Quantum Fluids* (Proceedings of the Batsheva seminar, Haifa,

1968), New York: Gordon and Breach.

Wojtaszek, Z., 1974, "The first years of cryogenics in the light of Olszewski's correspondence", *International Congress of the History of Science, XIII, 1971, Actes*, 7, 135—142.

Wolfke, M., Keesom, W. H., 1927, "On the change of the dielectric constant of liquid helium with the temperature. Provisional measurements", *CPL*, 190a, 3—14.

——, 1928, "New measurements about the way in which the dielectric constant of liquid helium depends on the temperature", *CPL*, 192a, 3—10.

Zahar, E., 1973, "Why did Einstein programme supersede Lorentz's I", *British Journal for the Philosophy of Science*, **24**, 95—123.

Index